I0493693

Report of Investigations 9693

Impact on Respirable Dust Levels When Operating a Flooded-bed Scrubber in 20-foot Cuts

Jay F. Colinet, W. R. Reed, and J. Drew Potts

DEPARTMENT OF HEALTH AND HUMAN SERVICES
Centers for Disease Control and Prevention
National Institute for Occupational Safety and Health
Office of Mine Safety and Health Research
Pittsburgh, PA • Spokane, WA

December 2013

Disclaimer

Mention of any company or product does not constitute endorsement by the National Institute for Occupational Safety and Health (NIOSH). In addition, citations to Web sites external to NIOSH do not constitute NIOSH endorsement of the sponsoring organizations or their programs or products. Furthermore, NIOSH is not responsible for the content of these Web sites. All Web addresses referenced in this document were accessible as of the publication date.

Ordering Information

To receive documents or other information about occupational safety and health topics, contact NIOSH at

>   Telephone: **1–800–CDC–INFO** (1–800–232–4636)
>   TTY: 1–888–232–6348
>   CDC INFO: **www.cdc.gov/info**
>
>   or visit the NIOSH Web site at **www.cdc.gov/niosh**.

For a monthly update on news at NIOSH, subscribe to NIOSH *eNews* by visiting **www.cdc.gov/niosh/eNews**.

Suggested Citation

NIOSH [2013]. Impact on Respirable Dust Levels When Operating a Flooded-bed Scrubber in 20-foot Cuts. By Colinet JF, Reed WR, Potts JD. Pittsburgh, PA: U.S. Department of Health and Human Services, Centers for Disease Control and Prevention, National Institute for Occupational Safety and Health, DHHS (NIOSH) Publication 2014–105, RI 9693.

DHHS (NIOSH) Publication No. 2014–105

December 2013

SAFER • HEALTHIER • PEOPLE™

Cover photo by NIOSH.

Contents

Figures

Tables

ACRONYMS AND ABBREVIATIONS

adj	adjusted
avg	average
CFR	Code of Federal Regulations
CM	continuous miner
conc	concentration
CWP	coal workers' pneumoconiosis
FBS	flooded-bed scrubber
int	intake
MMU	mechanized mining unit
MRE	Mining Research Establishment
MSA	Mine Safety Appliances
MSHA	Mine Safety and Health Administration
na	not available
NIOSH	National Institute for Occupational Safety and Health
No.	number
OMSHR	Office of Mine Safety and Health Research
oper	operator
PDM	personal dust monitor
pDR	personal DataRAM
PEL	permissible exposure limit
PVC	polyvinyl chloride
ret	return
RB	roof bolter
RRC	right rear corner
SC	shuttle car
SD	standard deviation

UNIT OF MEASURE ABBREVIATIONS

cfm	cubic feet per minute
ft	foot
in	inches
in Hg	inches of mercury
l/min	liters per minute
mg	milligrams
mg/m^3	milligrams per cubic meter
min	minute
mm	millimeter
%	percent
psi	pounds per square inch
μg	Microgram
μg/m^3	micrograms per cubic meter

Impact on Respirable Dust Levels When Operating a Flooded-bed Scrubber in 20-foot Cuts

Jay F. Colinet,[1] W. R. Reed,[2] and J. Drew Potts[3]

Abstract

Underground coal mining companies that operate continuous miner sections often apply to the Mine Safety and Health Administration (MSHA) for approval to take extended cuts to depths of up to 40 ft as a means of improving productivity. Historically, MSHA has granted approval of this practice if the mine has successfully demonstrated the ability to control the roof, methane, and respirable dust while extracting these extended cuts. A key component for controlling dust generated by continuous miners in 40-ft cuts has been the utilization of flooded-bed scrubbers. These fan-powered scrubbers pull dust-laden air from the mining face and remove respirable dust particles by passing the collected air through a wetted filter panel. The filtered air is then discharged back into the mine atmosphere. To effectively use scrubbers in faces that employ exhaust ventilation, the return ventilation curtain or tubing should be located outby the scrubber discharge on the continuous miner, which results in a setback distance from the face of approximately 40 ft. Over the last several years, MSHA has emphasized that mines demonstrate effective dust control before granting approvals for taking extended cuts with extended curtain setbacks. Each mine operator must successfully demonstrate control of workers' dust exposures in standard 20-ft cuts before MSHA considers approving extended cuts.

The goal of the research conducted by the National Institute for Occupational Safety and Health (NIOSH) was to compare dust levels generated in 20-ft cuts when using traditional exhaust face ventilation without a scrubber to dust levels in 20-ft cuts when using extended curtain setbacks with a scrubber operating. Dust surveys were completed at three mines, with area and personal sampling conducted to quantify respirable dust concentrations on a cut-by-cut basis. Dust sampling results did not show a statistically significant difference (Wilcoxon two-sample test, $\alpha = 0.05$) in respirable dust concentrations between these two test conditions (scrubber-on and scrubber-off) at the continuous miner or shuttle car sampling locations at the face. However, with the scrubber operating, respirable dust concentrations in the return airstream downwind of the continuous miner showed reductions of 91%, 86%, and 40% at Mines A, B, and C, respectively. The reductions at Mines A and B were found to be statistically significant when using the Wilcoxon test. Likewise, reductions in respirable quartz dust levels in the continuous miner return were observed at all three mines, with statistically significant reductions of over 80% observed at Mines A and B. Although operation of the flooded-bed scrubber did not impact respirable dust levels in the face area, it did significantly reduce respirable and quartz dust levels downwind of the continuous miner. Consequently, operation of the flooded-bed scrubbers, in conjunction with the dust controls required in the MSHA-approved ventilation plans at these mines, was advantageous from a respirable dust control perspective.

---

[1] Senior Scientist; Dust, Ventilation, and Toxic Substances Branch; Office of Mine Safety and Health Research (OMSHR); NIOSH, Pittsburgh, PA.
[2] Mining Engineer; Dust, Ventilation, and Toxic Substances Branch; OMSHR; NIOSH, Pittsburgh, PA.
[3] Branch Chief; Dust, Ventilation, and Toxic Substances Branch; OMSHR; NIOSH, Pittsburgh, PA

Introduction

During the mining and transport of coal, respirable dust is generated and can be liberated into the air ventilating the mine. If mine workers inhale excess amounts of this respirable airborne dust, it can result in the development of coal workers' pneumoconiosis (CWP). Likewise, if rock within or adjacent to the coal seam is extracted, respirable silica (quartz) dust can be liberated into the ventilating air. Inhalation of excess amounts of silica can lead to the development of silicosis. CWP and silicosis are disabling and potentially fatal lung diseases that cannot be cured [Cohen and Velho 2002], so it is critical to prevent mine workers from contracting these lung diseases.

The Federal Coal Mine Health and Safety Act of 1969 (Public Law 91-173) established a respirable coal mine dust standard of 2 mg/m^3 over an eight-hour shift. This Act also required periodic occupational dust sampling with a personal gravimetric sampler on each mechanized mining unit (MMU) to demonstrate compliance with the dust standard. When the quartz content in an MSHA-collected respirable dust sample exceeds 5%, a reduced dust standard is calculated by dividing the percent quartz into 10. For example, if a sample contains 11% quartz, the reduced dust standard for that MMU would be 0.9 mg/m^3 (10 ÷ 11% quartz = 0.9).

In an effort to limit the respirable dust exposure of coal mine workers, MSHA requires each mine operator to establish an acceptable ventilation plan that details how the mine proposes to control worker respirable dust exposure by defining ventilation parameters and dust control technologies that will be used [30 CFR 75.371][*]. This plan must be submitted to the appropriate MSHA district manager and approved by MSHA before mining can begin [30 CFR 75.370]. After the plan is approved, the following MSHA requirements apply:

> A person designated by the operator shall conduct an examination to assure compliance with the respirable dust control parameters specified in the mine ventilation plan. In those instances when a shift change is accomplished without an interruption in production on a section, the examination shall be made anytime within 1 hour of the shift change. In those instances when there is an interruption in production during the shift change, the examination shall be made before production begins on a section. Deficiencies in dust controls shall be corrected before production begins or resumes. The examination shall include air quantities and velocities, water pressures and flow rates, excessive leakage in the water delivery system, water spray numbers and orientations, section ventilation and control device placement, and any other dust suppression measures required by the ventilation plan [30 CFR 75.362 (a)(2)].

In addition to the above requirements, MSHA personnel assess the status of the dust controls during inspections to ensure that the plan parameters control respirable dust levels to or below the applicable dust standard. Typically, the mine will be permitted to exceed the minimum specified plan parameters by 120% and still be considered to be operating under the existing dust control plan [MSHA 2012a].

---

[*]Code of Federal Regulations. See CFR in references.

In order to extract extended cuts (i.e., those that extend beyond 20 ft), mine operators must apply to MSHA for approval. MSHA does not grant approval until after completing an on-site evaluation of the proposed extended cut system. Initially, MSHA will only allow for extraction of extended cuts while MSHA personnel are present at the mine site conducting their evaluation. At all other times, the mine operator must follow the standard cut plan included in the MSHA-approved ventilation plan, which "should require that the ventilation curtain/tubing be maintained to not more than 20 ft from the deepest point of penetration the face has been advanced" [MSHA 2012b].

In underground coal mines in the United States, the majority of continuous mining machines in use today are equipped with a flooded-bed scrubber, which is a fan-powered dust collector. The scrubber pulls dust-laden air from the face and passes it through a wetted filter panel. Past research by Colinet et al. [USBM 1990] has shown that scrubbers can remove over 90% of the respirable dust that is pulled into the unit. Although MSHA does not consider the scrubber to be a ventilation device, it has been shown to assist in moving ventilating air to the face [Taylor et al. 1996]. As a result of these dust collection and air-moving capabilities, flooded-bed scrubbers have become a key component for mines in receiving MSHA approval to take extended cuts of up to 40 ft as part of their approved ventilation plan.

In order to realize the maximum effectiveness of scrubbers in mines using exhausting face ventilation, the mouth of the brattice curtain should be outby the discharge of the scrubber so that the scrubber exhaust can be directed into the return airway [Jayaraman et al. 1990]. This would require the curtain to be approximately 40 ft from the face. In extended cut approvals, MSHA grants a variance to the mining operation to allow for this extended curtain setback distance. Without this variance, flooded-bed scrubbers cannot be effectively used in faces with exhaust ventilation. Historically, mines have not requested and MSHA has not granted this curtain setback variance for use in 20-ft cuts. However, the use of a flooded-bed scrubber in 20-ft cuts along with an extended curtain setback could be advantageous for controlling dust.

Researchers from the NIOSH Office of Mine Safety and Health Research (OMSHR), in partnership with MSHA and Alpha Natural Resources, completed a research project to evaluate the impact on respirable dust levels generated in 20-ft cuts with and without a flooded-bed scrubber operating. An extended curtain setback was used in cuts when the scrubber was operated. Respirable dust surveys were conducted at three Alpha underground coal mines (referred to here as Mines A, B, and C). Two of these mines did not have approval to operate a flooded-bed scrubber in their current ventilation plans. For these mines, mine management submitted a temporary plan to MSHA, which was approved by MSHA for use only during the NIOSH testing period. The third mine had the capability of operating a flooded-bed scrubber with an extended curtain setback to extract 40-ft cuts in its current plan, so no temporary plan was needed. However, the mine restricted the depth of the cuts taken during the NIOSH sampling to 20 ft to accommodate the test program parameters.

Prior to NIOSH initiating each of its dust surveys at the individual mines, MSHA personnel visited the mines to ensure that the plan parameters specified in the approved ventilation plans for Mines A, B, and C were in use. MSHA also examined the scrubbers to document their proper operation and quantify the air quantities produced. Table 1 summarizes a portion of the plan parameter information gathered by MSHA during the baseline surveys.

MSHA also conducted dust sampling during these visits to demonstrate that the approved plan parameters were successful in maintaining respirable dust at or below allowable levels. Under the operating conditions shown in Table 1, dust levels measured by MSHA over an 8-hour sampling period during the baseline surveys were within allowable levels. In addition to these baseline surveys, an MSHA representative participated in the subsequent NIOSH dust surveys at each mine and verified that minimum plan parameters were present during the NIOSH sampling.

Table 1. Minimum plan parameters and levels measured by MSHA in baseline surveys

Ventilation plan parameter	Mine A	Mine B	Mine C
Minimum face airflow, cfm	7,000	7,000	5,000
MSHA-measured face airflow, cfm	8,792	9,068	10,612[*]
Minimum scrubber airflow, cfm	6,400	6,000	4,000
MSHA-measured scrubber airflow, cfm	7,067	7,604	4,160
Minimum number of sprays operating at start of each cut	51	42	33
Number of sprays found operating at start of each cut	51	42	33
Minimum spray operating pressure, psi	65	65	85
MSHA-measured spray operating pressure, psi	74	74	92

[*]Average = 7,218 cfm when airflow quantities from three flush cuts are omitted from calculation.

Mine Sites Tested

Each of the mines included in these surveys used line brattice to implement an exhausting face ventilation system. For the cuts with the scrubber off, the ventilation curtain was set back from the face a maximum of 20 ft. For the cuts with the scrubber operating, the ventilation curtain was set back a distance of up to 40 ft.

All of the continuous miners sampled were located on super sections—i.e., where two sets of mining equipment operate simultaneously within the same working section, and each set is ventilated by a separate split of intake air—with the sampled mining machine responsible for completing the development of entries on one side of the super section. Two shuttle cars were used to transport coal from the face to the section feeder. A twin-boom bolter was used to complete the installation of roof bolts. In the approved ventilation plans for each of the MMUs sampled, the bolting machine was permitted to travel downwind of the continuous miner once during each production shift.

Appendices A, B, and C contain a diagram of the sections sampled at Mines A, B, and C, respectively. These appendices also contain additional information specific to each mine site. Table 2 provides a brief summary of the equipment and operating conditions at each mine during the NIOSH dust surveys.

Table 2. Summary of conditions observed during sampling at three mines

Mine parameter	Mine A	Mine B	Mine C
Continuous miner model	Joy 14 CM 15	Joy 12 CM 15	Joy 14 CM 15
Average mining height, in	62	94	65
Average face airflow, cfm	8,300	7,943	6,565
Average scrubber airflow, cfm	6,400	7,218	4,288
Scrubber filter panel density	30-layer	30-layer	30-layer
Number of water sprays operating	51	42	33
Spray type	BD3-5 hollow cone	BD3 hollow cone	BD3-5 hollow cone
Average water pressure, psi	99	68	100
Roof bolter model	Fletcher Roof Ranger II	Fletcher DDR-13	Fletcher Roof Ranger II
Average vacuum pressure, in Hg—left	16	14	15
Average vacuum pressure, in Hg—right	13	17	13

As noted in Table 2, all three mining machines used 30-layer filter panels in their respective flooded-bed scrubbers; however, the other parameters showed substantial variation across the three mines. For example, mining height varied by over 30 inches, while scrubber airflow varied by nearly 3,000 cfm. Consequently, the mines sampled during these surveys represent a range of operating conditions.

Sampling Protocol

Respirable dust measurements were made using the following instruments: (1) Thermo Scientific Model pDR-1000AN Personal Data Rams (pDRs); (2) Thermo Scientific Model PDM3600 Personal Dust Monitor (PDM); and (3) Mine Safety Appliances (MSA) Escort ELF sampling pumps (gravimetric samplers). The pDR samplers use light-scattering technology to provide real-time measurements of dust levels, which are stored in an internal data logger. Dust measurements were recorded by each pDR at five-second intervals throughout the sampling shift. The PDM sampler is a mass-based, real-time dust sampler that was operated at 2.2 l/min with a Higgins-Dewell cyclone. The PDM sampler recorded dust measurements at one-minute sampling intervals. Each MSA pump was operated at 2 l/min while connected to a Dorr-Oliver 10-mm nylon cyclone fitted with a 37-mm-diameter PVC filter. All filters were pre- and post-weighed at the OMSHR Pittsburgh lab, with the gain in dust mass and sampling time used to calculate the average respirable dust concentration for the sampling period. These time-weighted average dust concentrations were not converted to Mining Research Establishment (MRE) equivalent concentrations as described in 30 CFR 70.206 because they were not collected as compliance dust samples.

Because light-scattering instruments can be impacted by changes in dust composition and size distribution, the pDR manufacturer recommends that individual dust readings be corrected with a ratio calculated from adjacent gravimetric samplers [Thermo Scientific 2008]. Consequently, two gravimetric samplers were located side-by-side with a pDR sampler on area sampling racks. The average gravimetric concentration from the two samplers was divided by the pDR concentration for the entire sampling period to calculate a correction factor. The individual instantaneous dust readings from the pDR were then multiplied by this correction factor. The corrected pDR concentrations were used in all subsequent calculations.

Area sampling packages as described above were placed at multiple sampling locations, as illustrated in Figure 1. In addition, a gravimetric-only sampling package was operated in the miner return, with the filters from this sampling used for quartz analysis. A PDM sampler was worn by the continuous miner operator to assess his dust exposure. Each of these sampling locations will be discussed below in greater detail.

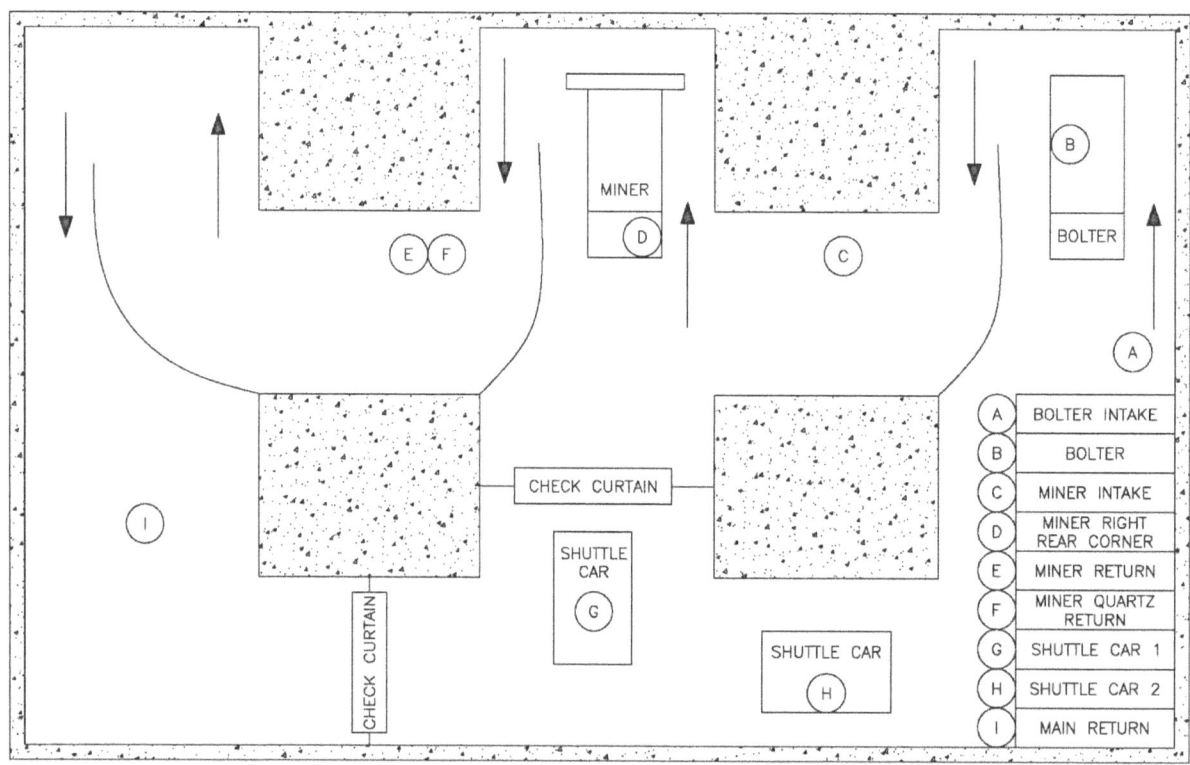

Figure 1. Schematic of typical dust sampling locations at each mine.

NIOSH researchers collected time study information related to the operation of the continuous miner. The start and stop times for each miner cut were used to calculate the average dust concentration during each cut. The times when each shuttle car entered and exited the active face were also recorded and used to calculate dust levels at each shuttle car while being loaded. When the continuous miner was not loading for periods of three min or longer, these down-periods were removed from the dust calculations so that the reported data would only represent dust generated during mining and loading.

Isolation of miner-generated dust was accomplished by placing area sampling packages in both the immediate intake and return (locations C and E in Figure 1) for the continuous miner. Intake dust levels were subtracted from return dust levels to calculate the quantity of dust attributable to the miner. An area sampling package was also located on the right rear corner (RRC) of the continuous miner (location D). This sampling location assisted in monitoring dust

6

rollback from the face. After the survey at Mine A was completed, an additional area sampling package was placed in the main return entry (location I) to expand the information obtained on differences in dust levels downwind of the miner. Figure 2 shows representative area sampling packages that were placed in the miner intake and on the rear corner of the continuous miner for each survey.

Photo by NIOSH   Photo by NIOSH

Figure 2. Dust sampling packages located in the intake to the continuous miner (left) and on the right rear corner of the miner (right).

An area sampling package was placed in each shuttle car cab (locations G and H) to monitor face area dust concentrations. The shuttle car dust concentrations are important for this study because their position with respect to the mining machine is consistent during mining and loading operations at the face. These data allow for a direct comparison of dust concentrations in the face area for the scrubber-on and scrubber-off test conditions.

The roof bolting machine was operated downwind of the continuous miner on several occasions during the survey. Because miner-generated dust can significantly impact the exposure of downwind personnel [Jayaraman et al. 1988], NIOSH monitored dust concentrations with sampling packages placed in the bolter intake (location A) and on-board the bolter (location B). Because of differences in operating practices and bolter designs, the bolter-mounted sampling package could not be placed at identical locations for all three surveys. A detailed time study documented the bolter's location and position with respect to the continuous miner (upwind or downwind) for each bolting cycle. Figure 3 illustrates representative locations for the sampling packages on the shuttle cars and roof bolters.

Miner-generated dust was analyzed for quartz content for both the scrubber-on and scrubber-off conditions. This was accomplished by using two sets of cyclones/filters on the same area sampling rack, which was placed in the immediate miner return (location F). If the scrubber was being used (scrubber-on), the hoses for two filters were connected to the sampling pumps and the pumps were activated just before mining commenced. The pumps were placed on hold at the end of the cut and the hoses were removed from the pumps. A similar procedure was followed with the other two filters during scrubber-off cuts. Sampling times were recorded for both conditions to allow for the calculation of airborne concentration, as well as percentage of quartz in the sample. The quartz content of these filters was determined by the MSHA analytical lab in Pittsburgh through the use of the MSHA P7 analytical method [MSHA 1989].

Photo by NIOSH   Photo by NIOSH

Figure 3. Dust sampling packages located on a shuttle car (left) and on a roof bolter (right).

As found in research by Potts et al. [NIOSH 2011], during normal mining operations, a buildup of material can occur in flooded-bed scrubbers, diminishing their air-moving effectiveness. This possibility was monitored by using a pitot tube and differential pressure gauge to measure velocity head in order to calculate scrubber air quantity prior to the start of each 20-ft scrubber-on cut configuration. At the beginning of each mine survey, a full traverse of velocity measurements (15 to 21 points depending upon the cross-sectional area of the scrubber ductwork) was completed and used to calculate scrubber air quantity. During subsequent scrubber measurements, three center-line velocity head measurements were obtained and compared to the three center-line readings from the full traverse to ensure appropriate airflow through the scrubber prior to each cut. As specified in the ventilation plans, the 30-layer scrubber screen was removed and cleaned with water after each cut. Additional maintenance included cleaning the scrubber inlets and ductwork at the beginning of each shift. In a few instances, the demister also had to be removed and cleaned in order to obtain the required scrubber airflow.

Air velocities in the entries and behind the exhaust line curtains were measured with a vane anemometer. An airflow measurement was taken at the face-side end of the ventilating curtain prior to the start of each cut. For cuts where the scrubber was going to be operated, this airflow measurement was taken before activation of the scrubber. Curtain configuration and length were also recorded. The exhaust brattice curtain was positioned just outby the scrubber exhaust for the scrubber-on tests and within 20 ft of the face during the scrubber-off tests. The scrubber discharged airflow either from the left or from the right rear corner of the miner, with that choice dictated at each mine by the location of the exhaust line brattice within the entry.

Past research has shown that the mining machine's external water spray system can have a dramatic impact on dust levels at the face [Organiscak and Beck 2010]. Therefore, the number of sprays operating and the operating pressure on the left and right side of the cutter head were monitored throughout the survey for adherence to the specifications in the ventilation plan for each MMU. Similarly, vacuum measurements were taken on the left and right drill heads of the roof bolting machine to ensure that the bolter dust collector was meeting the required operating levels.

Data Analysis

Detailed data on the cut locations, cutting times, face airflows, scrubber airflows, gravimetric, and real-time sampling dust concentrations for Mines A, B, and C are provided in the mine-specific survey reports in Appendices A, B, and C, respectively. In order to evaluate the impact of the flooded-bed scrubber in 20-ft cuts, the real-time sampling data and time study information were used to calculate dust concentrations on a cut-by-cut basis. As shown in the appendices, seven cuts were obtained for each of the scrubber-off and scrubber-on test conditions at Mines A and B, and four cuts were obtained for each of the test conditions at Mine C. Given the small sample sizes and the possibility that the dust data would not be normally distributed, it was necessary to use a nonparametric or distribution-free statistical test to analyze for differences between mean dust levels for each test condition.

The Wilcoxon[1] two-sample test (exact form) was used with a level of significance of $\alpha = 0.05$ to test the null hypothesis that the mean ranks of the two groups of dust data from the two test conditions were equal. Because the dust levels with the scrubber operating could be higher or lower than with the scrubber off, a two-tailed test was selected to identify any statistically significant differences between the two test conditions. If the computed probability (p-value) from the statistical test was less than 0.05, the observed difference between the mean ranks for the two test conditions was considered to be statistically significant. The average dust concentrations from the individual cuts, as shown in the tables in the appendices, were used to analyze the data for each sampling location at each mine. The mean dust concentrations for the scrubber-off and scrubber-on test conditions at each mine and the calculated p-values from the Wilcoxon test are provided in Tables 3, 4, and 6 to identify statistical significance.

Continuous Miner Results

The miner operator dust concentrations were calculated from the PDM data, and dust concentrations from all other sampling locations were calculated from pDR data. The dust concentrations for the miner operator, RRC, immediate miner return, and main return sampling locations had the intake dust levels subtracted out. Because increases in production can generate more dust [Webster et al. 1990] and increases in airflow can dilute dust levels [Hartmann 1973], the intake-adjusted dust levels were then normalized for differences in production and face airflow between cuts, assuming a linear relationship between these parameters and dust levels. For example, if average productivity for all cuts sampled during a mine survey was 0.28 shuttle cars per min and productivity for an individual cut averaged 0.36 shuttle cars per min, then the average dust level for this cut was multiplied by a factor of 0.78 (0.28 ÷ 0.36) to adjust for the higher-than-average productivity observed in the cut. The adjustment for differences in airflow were made in a similar manner, except that the factor was calculated by dividing the measured face airflow in an individual cut by the average face airflow of all cuts from the mine survey.

---

[1] The Wilcoxon two-sample test is the nonparametric counterpart of the independent-samples t-test. Whereas the t-test analyzes original data values, the Wilcoxon test analyzes rank-transformed data values. Ranks represent relative position within an ordered set of data values. The Wilcoxon test involves a comparison of the mean or average ranks of the two groups, just as the t-test involves a comparison of the means of original data values.

The airflow normalization relationship is the inverse of the productivity relationship since the dust concentrations are expected to decrease with an increase in face air quantity. A summary of the average normalized respirable dust concentrations around the continuous miner and in the return from the three surveys is provided in Table 3.

Table 3. Average respirable dust concentrations around the continuous miner and in the section return

Mine	Sampling location	Avg dust with scrubber off, mg/m^3	Avg dust with scrubber on, mg/m^3	Calculated p-value	Dust reduction with scrubber on
A	CM* intake	0.08	0.07	-	-
	CM operator	0.30	0.26	0.3310	13%
	CM RRC*	1.27	1.02	0.6439	20%
	CM return	23.86	2.05	0.0005[†]	91%
	Section return	na[‡]	na[‡]	na[‡]	na[‡]
B	CM intake	0.12	0.13	-	-
	CM operator	0.55	0.46	0.6352	16%
	CM RRC	7.53	5.20	0.6200	31%
	CM return	12.13	1.66	0.0005[†]	86%
	Section return	4.81	0.74	0.0005[†]	85%
C	CM intake	0.09	0.05	-	-
	CM operator	0.25	0.32	0.8000	-28%
	CM RRC	na[§]	na[§]	na[§]	na[§]
	CM return	14.37	8.64	1.000	40%
	Section return	6.73	4.80	0.400	29%

* Abbreviations: CM, continuous miner; RRC, right rear corner.
[†] Statistically significant difference.
[‡] No data available; section return sampling location added after completion of survey at Mine A.
[§] Water exposure of dust samplers invalidated data at this location for this mine.

Intake dust concentrations to the continuous miner faces were well controlled at all three mines and averaged 0.09 mg/m^3. Dust concentrations at the miner operator were 0.55 mg/m^3 or lower and exhibited a maximum difference of 0.09 mg/m^3 between scrubber-on and scrubber-off test conditions. These differences were not statistically significant. Dust concentrations at the RRC of the mining machine were somewhat variable from mine to mine, with no statistically significant difference despite the lower levels observed with the scrubber operating. However, major reductions in dust concentrations were observed in the immediate miner return when the scrubber was being operated. The reductions in average dust concentrations in the miner return at Mine A (21.81 mg/m^3) and at Mine B (10.47 mg/m^3) were statistically significant when evaluated using the Wilcoxon two-sample test with $\alpha = 0.05$. At Mine C, difficult geologic mining conditions resulted in the fewest cuts sampled for any of the surveys. In addition, the variability in measured dust concentrations observed between the cuts was greater than that observed at Mines A and B. When combining these two factors, the 5.73 mg/m^3 reduction in average miner return dust concentration with the scrubber operating at Mine C was not statistically significant.

For the average dust concentrations measured in the section return at Mine B, the 4.07 mg/m^3 reduction in dust concentration with the scrubber operating was statistically significant. A reduction of 1.93 mg/m^3 in the average section return dust concentrations was measured at Mine C, but this difference was not statistically significant due to the sampling issues previously identified for Mine C.

Shuttle Car Results

Time study data were collected to identify when each shuttle car entered the mining face to be loaded and when the car left the face. These time periods were used to calculate dust concentrations for each shuttle car while being loaded at the face. The individual dust concentrations were then used to calculate an average shuttle car dust concentration for each cut. The miner intake dust level was subtracted from the shuttle car dust level for each cut. The cut concentrations were used to calculate the average dust level with the scrubber off and the scrubber on at each mine. These average dust concentrations are summarized in Table 4.

Table 4. Average respirable dust concentrations at shuttle car cabs

Mine	Scrubber-off, mg/m^3	Scrubber-on, mg/m^3	Difference with scrubber on, mg/m^3	Calculated p-value[*]
A	0.27	0.28	**-0.01**	0.9272
B	0.03	0.07	**-0.04**	0.4272
C	0.35	0.43	**-0.08**	0.5429

[*]No statistically significant differences.

As shown in Table 4, operation of the scrubber during these surveys did not have any practical impact on dust concentrations at the shuttle car cab locations. The largest observed difference between scrubber-on and scrubber-off was an increase of only 0.08 mg/m^3, which was not statistically significant. The observed dust levels were 0.43 mg/m^3 or lower, indicating that dust was not escaping the miner face and was not exposing the shuttle car operators to elevated levels under either test condition.

Roof Bolter Results

Time study information was collected on the roof bolter to identify when the bolter was operating and the relative position of the bolter with respect to the continuous miner. Average dust concentrations were calculated for each roof bolter place and are summarized in Table 5a for the bolter intake and Table 5b for the on-board sampling locations at each mine.

When the bolter was positioned upwind of the miner, the intake air delivered to the bolter contained low levels of respirable dust and was 0.12 mg/m^3 or lower at these mines. Because the bolter is only permitted downwind of the miner once per shift, only a limited amount of downwind data was obtained. At Mine A, the bolter was only operated downwind of the miner once during the entire survey, so no scrubber-on/scrubber-off comparison could be made. However, the data from Mines B and C illustrate substantial reductions in dust levels when the

11

bolter was downwind and the scrubber was operating on the miner. Reductions of 34%–85% were measured at the bolter intake sampling location, with reductions of 51%–83% observed on the bolter. Because of the limited amount of data, statistical analysis was not completed.

Table 5. Average respirable dust concentrations at the roof bolter

a. Roof bolter intake sampling location

Mine	RB[*] upwind of miner, mg/m^3	RB downwind of miner with scrubber off, mg/m^3	RB downwind of miner with scrubber on, mg/m^3	**Reduction with scrubber on**
A	0.04	na[†]	3.29	**na[†]**
B	0.12	5.97	0.92	**85%**
C	0.09	11.23	7.38	**34%**

[*] Abbreviation: RB, roof bolter.
[†] No data available.

b. On-board sampling location[‡]

Mine	RB upwind of miner, mg/m^3	RB downwind of miner with scrubber off, mg/m^3	RB downwind of miner with scrubber on, mg/m^3	**Reduction with scrubber on**
A	0.12	na[†]	3.62	**na[†]**
B	0.30	5.65	0.94	**83%**
C	0.16	12.26	6.06	**51%**

[†] No data available.
[‡] Sampling package located near return-side-operator at Mine A, in the center of the walk-through bolter at Mine B, and in the operator's cab at Mine C.

In addition to dust generated by the continuous miner, bolter operators can be exposed to dust generated from installing bolts. However, when comparing the concentrations measured at the bolter intake location when upwind of the miner to those measured on the bolter, increases in dust concentrations of only 0.07 to 0.18 mg/m^3 were observed, indicating that the vacuum collection systems on these bolters were effective in capturing bolter-generated dust. NIOSH personnel measured the vacuum pressure at each drill head multiple times during each shift. All of the measured vacuum readings provided in Table 2 exceeded the minimum vacuum pressure required by MSHA.

Quartz Results

A gravimetric sampling package was placed in the immediate miner return and collected dust samples that were analyzed for quartz content. The quartz analysis provided the mass of quartz on each filter, which allowed for calculation of the quartz percent and concentration. In order to address the differences in production rates for the various cuts, the quartz concentrations from each cut were normalized based upon the average number of shuttle cars per min of sampling time. These values were used to calculate an average quartz concentration for the scrubber-off and scrubber-on test conditions. Table 6 summarizes the average quartz data from samples collected at each of the three mines.

The percent quartz found in the samples with the scrubber on showed mixed trends in that the percent was higher in Mines A (0.3%) and C (2.3%) and lower in Mine B (-0.9%). However, operation of the scrubber lowered the quartz mass and adjusted concentrations in all three mines. Results from Mines A and B were very similar with reductions in mass and concentration greater than 80%, which were statistically significant with the Wilcoxon test. At Mine C, the highest quartz percentages of all mines were observed. However, the reductions in mass (55%) and concentration (14%) with the scrubber on were not as great at Mine C and these differences were not statistically significant.

Table 6. Quartz content in the immediate miner return with the scrubber off and on

Quartz content in gravimetric samples	Mine A	Mine B	Mine C
Quartz percent with scrubber off	6.8	3.7	14.3
Quartz percent with scrubber on	7.1	2.8	16.6
Quartz mass with scrubber off, μg	202	38	387
Quartz mass with scrubber on, μg	31	7	175
Reduction in mass with scrubber on	86%	82%	55%
Adjusted quartz concentration with scrubber off, μg/m^3	1,282	255	1,602
Adjusted quartz concentration with scrubber on, μg/m^3	177	45	1,385
Calculated p-value	*0.0022*[*]	*0.0022*[*]	*0.3939*
Reduction in quartz concentration with scrubber on	**86%**	**82%**	**14%**

[*] Statistically significant difference.

Summary

Respirable dust samples were collected from continuous miner sections at three underground coal mines to evaluate the impact of operating a flooded-bed scrubber in 20-ft cuts with exhaust face ventilation and an extended curtain setback. The dust levels observed with the scrubber operating were compared to dust levels measured in 20-ft cuts without a scrubber operating and the exhaust ventilation curtain advanced to within 20 ft of the face. Area dust sampling with gravimetric and light-scattering instrumentation was conducted in the intake and return airways of the continuous miner, on the continuous miner, in the shuttle car cabs, and at the roof bolting machine. A personal dust monitor (PDM) was worn by the continuous miner operator to measure the dust exposure at the operator's position. Time study information was collected by NIOSH researchers so that the dust data could be analyzed on a cut-by-cut basis, with extended downtimes removed from the calculations. MSHA personnel participated in the dust surveys and monitored operating parameters specified in the individual mine ventilation plans to ensure that minimum plan parameters were present during each survey. Three days of sampling were conducted at each of the mine sites. When sufficient data were available, the Wilcoxon two-sample test ($\alpha = 0.05$) was used to determine the statistical significance of differences in mean dust levels between the two test conditions.

Analysis of the data from the three surveys supports the following findings:

- Intake dust concentrations to the continuous miner faces were well controlled at all three mines and averaged just under 0.1 mg/m^3.

- There were no statistically significant differences in dust concentrations between test conditions for the continuous miner operator at each mine, with a maximum difference in average dust concentration of only 0.09 mg/m^3.

- With the scrubber operating, dust concentrations at the right rear corner of the miner were 22% lower at Mine A and 31% lower at Mine B. However, these reductions were not statistically significant.

- Dust concentrations in the return immediately downwind of the miner were at least 40% lower when the scrubber was operating. At Mines A and B, statistically significant reductions of 91% and 86% were observed, respectively.

- A sampling package was added in the section return for the surveys conducted at Mines B and C. A statistically significant reduction of 85% was observed in section return dust concentrations at Mine B with the scrubber operating, and a reduction of 29% was observed at Mine C.

- There were no statistically significant differences in dust concentrations measured at the shuttle cars between test conditions. The maximum difference measured at the shuttle cars between test conditions at any of the mines was 0.08 mg/m^3.

- When the roof bolter was operating downwind of the continuous miner, intake dust concentrations to the bolter were reduced by 85% at Mine B and 34% at Mine C when the scrubber was operated. Limited data for these conditions prevented statistical analysis. At Mine A, the bolter was only downwind of the miner one time during the survey, so no comparison could be made.

- Quartz mass and concentrations in the continuous miner return were reduced at all three mines when the scrubber was operating. At Mines A and B, statistically significant reductions of over 80% were measured.

Results from the surveys at these three mines show that there were no statistically significant differences in dust concentrations at the continuous miner operator, the right rear corner of the miner, or in the shuttle car cabs when the flooded-bed scrubber was operated. However, statistically significant reductions in respirable and quartz dust levels in the miner and section return were measured during the surveys. As a result, dust exposures for personnel located downwind of the miner would be reduced through operation of the flooded-bed scrubber, with no adverse impact to workers located in the face area. Therefore, from a dust control perspective, operation of the flooded-bed scrubber would be beneficial.

It should be noted that these results were achieved with the 30-layer scrubber filter panels being cleaned after each 20-ft cut, which was found to be necessary in order to maintain scrubber airflows at the required levels. To illustrate, airflow measurements were taken in the scrubber after completion of one 20-ft cut during which the scrubber was operated but before the filter panel had been cleaned. On three different occasions during the survey at Mine B, measurements were taken after completion of one cut with the scrubber operating. An average drop in scrubber airflow of 2,000 cfm was observed, which represented an average reduction of 29% when compared to the quantities available at the beginning of each of these cuts. At Mine C, a scrubber airflow reading was taken after one cut and before the filter panel was cleaned. A drop in airflow of 1,500 cfm was measured, which represented a reduction of 35%. These numbers emphasize the critical need to clean the 30-layer filter panel after each cut. A previous research study by Potts et al. [NIOSH 2011] recommended the cleaning of scrubber filter panels after every 40 ft of advance when using a 20-layer filter. In addition, the minimum operating parameters specified in the ventilation plans were present prior to the start of each cut and contributed to the results observed during these surveys.

Acknowledgements

The authors would like to acknowledge the invaluable access provided by Alpha Natural Resources to its mine sites and the assistance provided by the mining personnel at each of the mines. Also, the authors appreciate the effort and assistance provided by numerous MSHA personnel through input into the test protocol, completion of baseline surveys at each MMU, quartz analysis of filters at MSHA's Pittsburgh laboratory, and for their efforts in verifying plan parameters and providing additional assistance to NIOSH personnel during sampling at each of the mines. The authors express appreciation to Linda McWilliams of NIOSH for her guidance and assistance in completing the statistical analysis of the dust data. Finally, the authors would like to acknowledge the significant contributions of NIOSH technicians Jason Driscoll and Milan Yekich for their assistance during the mine surveys and in preparing the dust sampling equipment for each of these surveys.

References

CFR. Code of Federal Regulations. Washington, DC: U.S. Government Printing Office, Office of the Federal Register.

Cohen R, Velho V [2002]. Update on respiratory disease from coal mine and silica dust. Clinics in Chest Medicine 23(4):811–826.

Hartmann HL [1973]. Mine atmospheres and gases. In: Cummins AB, Given IA eds. SME Mining Engineering Handbook. Vol. 1. Baltimore: Port City Press, pp. 16-2–16-4.

Jayaraman NI, Babbitt C, and O'Green J [1988]. Ventilation and dust control techniques for personnel downwind of continuous miner. SME Transactions, Vol. 284, Littleton, CO: Society for Mining, Metallurgy, and Exploration, Inc., pp. 1823–1826.

Jayaraman NI, Volkwein JC, Kissell FN [1990]. Update on continuous miner dust scrubber applications. Mining Engineering 42(3):281–284.

MSHA [1989]. Determining the quartz content of respirable dust by FTIR. Arlington, VA: U.S. Department of Labor, Mine Safety and Health Administration, Information Report 1169.

MSHA [2012a]. Coal mine health inspection procedures. Arlington, VA: U.S. Department of Labor, Mine Safety and Health Administration, MSHA Handbook Series, Handbook Number: PH89-V-1(23).

MSHA [2012b]. Procedures for evaluation of requests to make extended cuts with remote controlled continuous mining machines. Arlington, VA: U.S. Department of Labor, Mine Safety and Health Administration, Procedure Instruction Letter NO. I12-V-11.

NIOSH [2011]. Evaluation of face dust concentrations at mines using deep-cutting practices. By: Potts JD, Reed WR, Colinet JF. Pittsburgh, PA: U.S. Department of Health and Human Services, National Institute for Occupational Safety and Health, DHHS (NIOSH) Publication No. 2011-131, RI 9680.

Organiscak JA, Beck TW [2010]. Continuous miner spray considerations for optimizing scrubber performance in exhaust ventilation systems. Mining Engineering 62(10): 41–46.

Taylor CD, Rider JP, Thimons ED [1996]. Changes in face methane concentrations using high capacity scrubbers with exhausting and blowing ventilation. Littleton, CO: Society for Mining, Metallurgy, and Exploration, Pre-print 96-167.

Thermo Scientific [2008]. Model pDR-1000AN/1200 instruction manual. Franklin, MA: Thermo Scientific, pp. 35–36.

USBM [1990]. Laboratory evaluation of quartz dust capture of irrigated-filter collection systems for continuous miners. By: Colinet JF, McClelland JJ, Erhard LA, Jankowski RA. Pittsburgh, PA: U.S. Department of the Interior, Bureau of Mines, RI 9313.

Webster JB, Chiaretta CW, Behling J [1990]. Dust control in high productivity mines. Littleton, CO: Society for Mining, Metallurgy, and Exploration, Pre-print 90-82.

Appendix A: Dust Survey at Mine A

Introduction

The purpose of this study was to compare face dust levels for two cutting conditions: (1) a standard 20-ft cut arrangement with no scrubber and a maximum 20-ft exhaust curtain setback, and (2) a 20-ft cut with exhaust curtain setbacks of up to 40 ft and operation of a flooded-bed scrubber. A three-day survey was conducted, alternating between both conditions on a cut-by-cut basis. This strategy results in minimizing the effects of changes in operating conditions from day to day. For operating conditions known to affect face dust concentrations—including face airflow and the number, orientation, type, location, and pressure of external water sprays—attempts were made to maintain these conditions in a consistent state throughout the study.

Site Description

This dust survey was conducted on the left side of a nine-entry super section, as shown in Figure A-1. Coal was mined using a Joy 14CM15 continuous mining machine and two Joy standard cab (off-curtain) shuttle cars. During the survey, entries 1 through 4 were developed with No. 3 and No. 4 entries used for intake air and No. 1 and No. 2 entries used as returns. The section beltline and feeder/breaker were located in No. 5 entry on a neutral air split. Mining height averaged 62 in, including the extraction of approximately 14 in of top rock. The pillars were on 60-ft centers with 20-ft entry widths, resulting in a 40-ft X 40-ft pillar dimension. The continuous miner faces were ventilated using exhausting curtain. Blowing or exhausting curtain can be used to ventilate the bolter faces and the mine uses a dual-head, Fletcher Roof Ranger II to install the bolts.

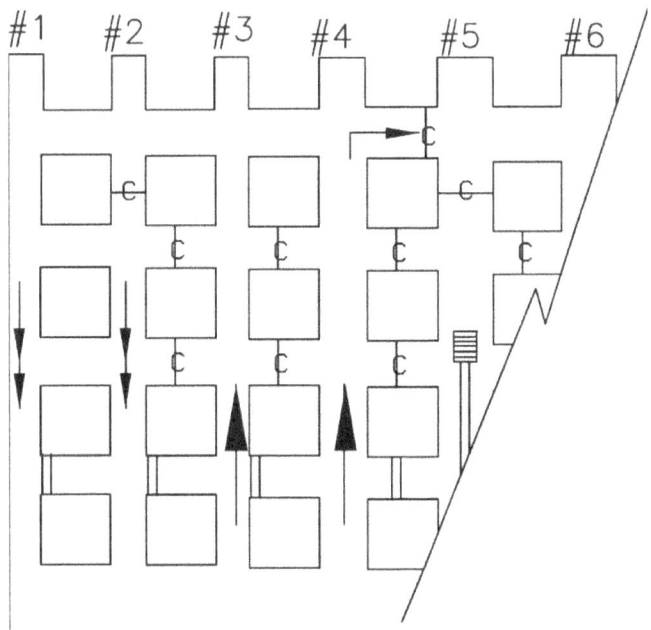

Figure A-1. Schematic of section sampled at Mine A.

Ventilation Plan Parameters

To assure normal operating conditions, the mine's adherence to the dust control parameters stipulated in the ventilation plan was monitored and corrective actions were taken when parameters fell outside required limits (less than the minimum specified or exceeding 120% of the minimum specified). The ventilation plan parameters used on this section for this study included the following minimum ventilation quantities: 15,000 cfm in the last open crosscut, 7,000 cfm at the face-side end of the brattice curtain on the continuous miner faces, 6,400 cfm scrubber airflow, and 4,000 cfm at the face-side end of the curtain on the bolter faces. A 30-layer filter panel was used in the scrubber. Maximum curtain setbacks were 20 ft for the nonscrubber cuts and 40 ft for the scrubber cuts. The continuous miner was equipped with 51 water sprays (3-5 Spraying System hollow cone), all of which had to be operational at the start of the cut and at least 40 had to be operational at all times. The minimum spray pressure was 65 psi.

Of the plan parameters monitored for the survey, curtain airflow on the continuous miner faces was the most difficult to maintain within 120% of the specified minimum level of 7,000 cfm. Based on this minimum, airflows between 7,000 and 8,400 cfm were targeted. The average curtain airflow during the survey was 8,300 cfm (SD = 670). At times it was difficult to keep airflows below 8,400 cfm, with 6 of 14 cuts being greater than this quantity; however, the airflows exhibited a fairly tight range of 2,200 cfm. The targeted scrubber airflow was between 6,400 and 7,680 cfm. Measured scrubber airflows averaged 6,400 (SD = 270), with only one reading falling outside the targeted range. The targeted curtain airflow on the bolting faces was between 4,000 and 4,800 cfm. Disregarding bolting activities in Room No. 9, which occurred in the last open crosscut, the average curtain airflow on the bolting faces was 4,400 cfm (SD = 370) and airflows on only 2 of 13 bolting faces fell outside of the targeted range. Airflow measurements in the last open crosscut, as well as water spray pressures, conformed well to the ventilation plan used during the survey. In general, mine personnel were able to adhere to the temporary ventilation plan approved for this dust survey.

Results

Table A-1 provides a summary of respirable dust concentrations around the mining machine, including miner intake, return, right rear corner, and operator locations, for each of the 14 cuts analyzed during the survey. Detailed data for each miner cut are provided in Table A-5 near the end of this appendix. The dual numbering system for cuts represents the shift number followed by the number of the cut extracted during that shift. Dust attributable to mining activity was calculated by subtracting the intake dust concentration from the measured dust concentration at each of the other sampling locations. The dust concentrations have been adjusted based on the measured productivity and airflow for each cut. The number of cars loaded per min was used as the measure of productivity and all dust levels were normalized to the average productivity observed throughout the study of 0.37 cars per min. For example, productivity during Cut No. 1-1 was 0.34 cars per min. Therefore, dust concentrations measured for this cut were normalized by multiplying by a correction factor of 1.09 (0.37/0.34) to adjust for the lower-than-average productivity. Likewise, average airflow during the survey was divided into the actual airflow measured for each cut to calculate the correction factor for normalizing the dust concentrations for differences in face airflow.

Table A-1. Summary of adjusted respirable dust concentrations for the continuous miner

a. Scrubber-off

Cut No.	Cut location	Scrubber airflow, cfm	Entry airflow, cfm	Start cut	End cut	SCs[*]/ min	CM[*] intake dust, mg/m^3	CM oper dust, mg/m^3	CM RRC[*] dust, mg/m^3	CM return dust, mg/m^3
1-1	4-heading	0	8,200	8:20	8:54	0.33	0.00	0.17	0.90	25.31
1-3	3-left	0	9,600	10:11	10:31	0.50	0.03	1.19	2.02	12.94
1-5	1-heading	0	8,600	12:18	12:42	0.34	0.01	0.21	0.10	28.15
2-2	4-heading	0	7,400	9:04	9:29	0.36	0.02	0.21	0.44	20.36
2-5	2-left	0	9,000	12:58	13:14	0.43	0.19	0.09	0.05	20.11
3-1	2-heading	0	8,000	8:21	8:38	0.30	0.11	0.00	1.67	34.97
3-3	1-left	0	7,700	11:47	12:09	0.41	0.23	0.20	3.69	25.21

[*] Abbreviations: SC, shuttle car; CM, continuous miner; RRC, right rear corner.

b. Scrubber-on

Cut No.	Cut location	Scrubber airflow, cfm	Entry airflow, cfm	Start cut	End cut	SCs/ min	CM intake dust, mg/m^3	CM oper dust, mg/m^3	CM RRC dust, mg/m^3	CM return dust, mg/m^3
1-2	3-heading	5,900	7,400	9:20	9:47	0.34	0.08	0.22	0.83	2.55
1-4	4-heading	6,600	7,700	11:14	11:45	0.41	0.01	0.39	0.44	2.03
2-1	1-heading	6,400	8,100	7:38	8:13	0.25	0.12	0.35	1.49	2.69
2-3	3-left	6,300	8,800	9:59	10:18	0.37	0.15	0.19	0.39	1.70
2-4	3-heading	6,700	8,800	11:32	12:25	0.34	1.05[†]	0.21	0.04	1.58
3-2	2-heading	6,500	7,900	9:44	11:14	0.44	0.02	0.00	0.64	1.75
3-4	3-heading	6,600	9,000	13:01	13:32	0.29	0.05	0.48	3.29	2.02

[†] Not included in calculation of average intake concentration.

c. Impact of scrubber performance

Test conditions and statistics	CM intake	CM operator	CM RRC	CM return
Average concentration with scrubber off, mg/m^3	0.08	0.30	1.27	23.86
Average concentration with scrubber on, mg/m^3	0.07	0.26	1.02	2.05
Calculated p-value for cut data	-	*0.3310*	*0.6439*	*0.0005[‡]*
Percent reduction in concentration with scrubber on	-	**13%**	**20%**	**91%**

[‡] Statistically significant difference

The intake dust level in Cut No. 2-4 was atypically high and was attributed to the intake pumps not properly positioned in the primary intake during this cut. Due to low dust levels (0.04 mg/m^3) at the RRC of the mining machine during Cut No. 2-4, intake dust levels were assumed to be 0.00 mg/m^3 for this cut. Average respirable dust concentrations were 91% lower in the return when using the scrubber, decreasing from 23.86 mg/m^3 (scrubber-off) to 2.05 mg/m^3 (scrubber-on). This difference was statistically significant when using the Wilcoxon two-sample test, with $\alpha = 0.05$. Average dust concentrations on the RRC of the mining machine averaged 1.01 mg/m^3 when using the scrubber versus 1.29 mg/m^3 when not using the scrubber; however, this difference was not statistically significant.

If continuously exposed to the average return dust concentration measured with the scrubber off (23.86 mg/m³) for only 30 min, a mine worker's respirable dust exposure would exceed the permissible exposure limit (PEL) of 2.0 mg/m³ (assumes conversion of the resulting concentration to an MRE equivalent). Consequently, use of the flooded-bed scrubber is recommended because a statistically significant reduction in return dust levels of over 90% was observed when the scrubber was operated.

Table A-1 also shows personal dust monitor (PDM) data collected at the continuous miner operator occupation. These data are corrected for intake dust concentrations as well as productivity. There is no statiscally significant difference in miner operator exposures when comparing the scrubber-on condition to the scrubber-off condition, averaging 0.26 mg/m³ and 0.30 mg/m³, respectively. The data collected around the continuous miner and in the shuttle car cabs indicate that the dust levels at the face were controlled when adhering to the requirements stipulated in the ventilation plan used during testing, regardless of scrubber use.

Table A-2 shows the average of the shuttle car dust data collected during the study for each cut while the cars were being loaded at the face. The individual dust concentrations measured for each shuttle car loading period with the scrubber off and on are provided at the end of the appendix in Tables A-6 and A-7, respectively. These data isolated the shuttle car dust exposures when located at the face during miner cutting and loading activities and do not include exposures when in transit or at the feeder/breaker. Intake dust concentrations were subtracted from shuttle car concentrations to arrive at adjusted shuttle car concentrations resulting from face activities. Shuttle car dust concentrations when loading at the face averaged 0.27 mg/m³ with the scrubber off, versus 0.28 mg/m³ with the scrubber on; however, this difference was not statistically significant (*p-value = 0.9272*).

Table A-2. Summary of adjusted respirable dust concentrations in the shuttle car cabs

Cut number with scrubber off	Cut location with scrubber off	SC* dust with scrubber off, mg/m³	Cut number with scrubber on	Cut location with scrubber on	SC dust with scrubber on, mg/m³
1-1	4-heading	0.52	1-2	3-Heading	0.15
1-3	3-left	0.37	1-4	4-Heading	0.64
1-5	1-heading	0.21	2-1	1-Heading	0.29
2-2	4-heading	0.34	2-3	3-Left	0.23
2-5	2-left	0.14	2-4	3-Heading	0.53
3-1	2-heading	0.03	3-2	2-Heading	0.09
3-3	1-left	0.29	3-4	3-Heading	0.01
Average	-	0.27	-	-	0.28

*Abbreviation: SC, shuttle car.

Table A-3 shows respirable dust concentrations in the vicinity of the return-side bolter operator and in the intake air split for the bolting machine. The intake levels were subtracted from the return-side levels to estimate bolter-generated dust. The bolting machine generated very little dust (approximately 0.10 mg/m³) when proper face ventilation was maintained (in this case 4,000 cfm) and the dust collector equipment was in good working order. The dust collector suction pressures averaged 15.7 in Hg (SD = 0.8) on the return-side of the bolter and 12.5 in Hg (SD = 0.5) on the intake-side. However, significant exposures to respirable dust can occur when the bolter machine operates downwind of the continuous mining machine. Bolter place No. 1-4 was bolted downwind of mining operations when the scrubber was in use. It was noted in the time study data that place No. 1-4 lost ventilating air from 11:27 to 11:32, causing dust levels to dramatically spike. Consequently, this time segment was removed from the calculation of dust levels. For place No. 1-4, respirable dust concentrations ranged from 3.29 mg/m^3 (intake) to 3.62 mg/m^3 (return-side operator) in the vicinity of the bolter when operating downwind of the miner under the scrubber-on condition. As shown in the miner return, these dust levels could be nearly a magnitude of order higher during scrubber-off conditions.

Table A-3. Summary of roof bolter dust concentrations

Place No.	Face location	Face airflow, cfm	Position of RB* with respect to CM*	Start time	End time	RB intake dust, mg/m^3	RB operator dust, mg/m^3	RB generated dust, mg/m^3
1-1	2-heading	4,400	upwind	7:53	8:08	0.08	0.01	0.00
1-2	4-heading	5,000	upwind	9:17	9:36	0.03	0.01	0.00
1-3	3-left	4,100	upwind	10:45	11:03	1.98†	0.22	void†
1-4	3-heading	4,400	downwind‡	11:29	11:44	3.29	3.62	0.33
1-5	4-heading	4,800	upwind	12:05	12:19	0.01	0.06	0.05
2-1	2-heading	4,300	upwind	7:12	7:30	0.05	0.31	0.26
2-2	1-heading	4,500	upwind	8:32	8:45	0.06	0.10	0.04
2-3	4-heading	4,100	upwind	9:50	10:04	0.05	0.12	0.07
2-4	3-left	10,000	upwind	10:25	10:45	0.05	0.10	0.05
2-5	3-heading	4,100	upwind	12:44	13:03	0.01	0.16	0.15
3-1	1-heading	4,100	upwind	7:21	7:28	0.05	0.09	0.04
3-2	2-heading	4,100	upwind	8:55	9:12	0.02	0.11	0.09
3-3	2-heading	4,000	upwind	11:36	11:54	0.05	0.15	0.10
3-4	1-left	5,100	upwind	12:30	12:47	0.00	0.08	0.07
Avg	-	-	-	-	-	-	-	**0.10**

* Abbreviations: RB, roof bolter; CM, continuous miner.

† Problem with intake sampler position voided dust reading and calculation of bolter-generated dust.

‡ Scrubber operating on miner while bolter was downwind.

Table A-4 displays the mass, percentage, and concentration of miner-generated respirable quartz for samples collected in the miner return for both the scrubber-off and scrubber-on conditions. The percentage of quartz in miner-generated dust was not affected by operation of the scrubber and averaged approximately 7%. However, use of the scrubber did reduce the measured mass of miner-generated quartz dust by 85% from 202 µg to 31 µg. Due to the difference in production between the test conditions, the quartz concentrations were normalized based upon the average number of shuttle cars per min of sampling time (0.24 cars/min). The adjusted quartz concentration showed a statistically significant difference with the Wilcoxon test and a reduction of 86% when the scrubber was operating. Based upon the quartz levels measured downwind of the miner with the scrubber off, an exposure to 1,282 µg/m^3 of quartz for only 30 min would result in a miner's exposure that exceeds 100 µg/m^3 (assumes conversion to an MRE equivalent). Consequently, operation of the scrubber is recommended in order to significantly reduce the quartz exposure for workers located downwind of the miner.

Table A-4. Respirable quartz levels in the continuous miner return

a. Scrubber-off

Shift	Gravimetric filter number	Sampling time, min	Dust mass, mg	Avg dust conc, mg/m³	Quartz mass, μg	Quartz, %	Quartz conc, μg/m³	Shuttle cars/ min	Adjusted quartz conc, μg/m³
1	567952	101	3.736	17.60	266	7.1	1,316.8	0.29	1,089.8
	566872	101	3.374		231	6.8	1,143.6	0.29	946.4
2	566870	58	2.796	23.16	199	7.1	1,715.5	0.28	1,470.5
	567944	58	2.578		175	6.8	1,508.6	0.28	1,293.1
3	567949	61	2.960	21.36	183	6.2	1,500.0	0.23	1,565.2
	566875	61	2.252		155	6.9	1,270.5	0.23	1,325.7

b. Scrubber-on

Shift	Gravimetric filter number	Sampling time, min	Dust mass, mg	Avg dust conc, mg/m³	Quartz mass, μg	Quartz, %	Quartz conc, μg/m³	Shuttle cars/ min	Adjusted quartz conc, μg/m³
1	566890	80	0.633	3.93	37	5.9	231.3	0.24	231.3
	566880	80	0.624		38	6.1	237.5	0.24	237.5
2	566883	126	0.384	1.49	31	8.1	123.0	0.21	140.6
	566881	126	0.369		32	8.7	127.0	0.21	145.1
3	566863	86	0.358	1.93	25	7.0	145.4	0.20	174.4
	566871	86	0.307		20	6.5	116.3	0.20	139.5

c. Impact of scrubber performance

Test conditions and statistics	Quartz mass, μg	Quartz conc, μg/m³	Adjusted quartz conc, μg/m³
Average quartz content with scrubber off	202	1,409.2	1,281.8
Average quartz content with scrubber on	31	163.4	178.1
Calculated p-value	-	-	0.0022[*]
Percent reduction in quartz content with scrubber on	**85%**	**88%**	**86%**

[*] Statistically significant difference.

Conclusions

The purpose of this study was to compare face dust levels for two cutting conditions: (1) a standard 20-ft cut advance without the scrubber operating and a maximum 20-ft curtain setback and (2) a 20-ft cut advance with the scrubber operating and a curtain setback of up to 40 ft. The mine's adherence to the dust control parameters stipulated in the temporary ventilation plan were monitored by MSHA personnel, and corrective actions were taken when parameters fell outside required limits (less than the minimum specified or exceeding 120% of the minimum specified).

In general, mine personnel were able to adhere to the levels specified in the temporary ventilation plan. A summary of the results follows:

- Average respirable dust concentrations were 91% lower in the return when using the scrubber, decreasing from 23.86 mg/m^3 to 2.05 mg/m^3. This difference is statistically significant.

- Dust concentrations on the RRC of the mining machine averaged 1.02 mg/m^3 when using the scrubber versus 1.27 mg/m^3 when not using the scrubber. However, this difference is not statistically significant.

- There is no statistically significant difference in miner operator exposures when comparing the scrubber-on condition to the scrubber-off condition, averaging 0.26 mg/m^3 and 0.30 mg/m^3, respectively.

- Shuttle car dust concentrations when loading at the face averaged 0.28 mg/m^3 with the scrubber versus 0.27 mg/m^3 without the scrubber. However, this difference is not statistically significant.

- The bolting machine generated very little dust (approximately 0.10 mg/m^3) when proper face ventilation was maintained (in this case 4,000 cfm), and the dust collector was in good working order.

- For one bolting cycle, respirable dust concentrations ranged from 3.29 mg/m^3 to 3.62 mg/m^3 in the vicinity of the bolter when operating downwind of the miner under the scrubber-on condition.

- The percentage of quartz in miner-generated dust was not affected by operation of the scrubber and averaged approximately 7%. However, use of the scrubber reduced the amount of miner-generated quartz dust by 86% from 1,282 μg/m^3 to 177 μg/m^3, and this difference is statistically significant.

The data collected for this study indicate that the mine was able to control dust levels in the vicinity of the mining machine, regardless of scrubber use, when adhering to the requirements stipulated in the temporary ventilation plan used during this survey. However, operation of the scrubber resulted in statistically significant reductions of respirable coal and quartz dust downwind of the miner. Without the scrubber operating, a short exposure downwind of the mining machine of 30 min (assumes average dust concentrations measured during the study and converted to an MRE equivalent) would result in an overexposure to respirable coal mine dust (>2.0 mg/m^3) and respirable quartz dust (>100 μg/m^3).

Table A-5. Continuous-miner-generated dust concentrations (mg/m³) for each cut

Cut No.	Cut location	Face air-flow, cfm	FBS* air-flow, cfm	CM* start time	CM end time	Total cut time, min	No. SC*	SCs/ min	CM int dust conc	CM oper dust conc	CM RRC* dust conc	CM ret dust conc	Adj† CM oper dust conc	Adj† CM RRC dust conc	Adj† CM ret dust conc
1-1	4-heading	8,200	0	8:20	8:54	34	11	0.33	0.00	0.16	0.81	22.85	0.17	0.91	25.62
1-2	3-heading	7,400	5,900	9:20	9:47	27	9	0.34	0.08	0.29	0.94	2.71	0.22	0.94	2.86
1-3	3-left	9,600	0	10:11	10:31	20	10	0.50	0.03	1.64	2.39	15.15	1.19	1.74	11.19
1-4	4-heading	7,700	6,600	11:14 11:33	11:26 11:45	24	10	0.41	0.01	0.44	0.53	2.44	0.39	0.47	2.19
1-5	1-heading	8,600	0	12:18	12:42	24	8	0.34	0.01	0.20	0.10	24.98	0.21	0.10	27.17
2-1	1-heading	8,100	6,400	7:38	8:13	35	9	0.25	0.12	0.36	1.15	1.98	0.35	1.53	2.76
2-2	4-heading	7,400	0	9:04	9:29	25	9	0.36	0.02	0.23	0.50	22.24	0.21	0.50	22.84
2-3	3-left	8,800	6,300	9:59	10:18	19	7	0.37	0.15	0.34	0.52	1.75	0.19	0.37	1.60
2-4	3-heading	8,800	6,700	11:32 12:06	11:41 12:25	28	10	0.34	1.05	0.19	0.04	1.37	0.21	0.04	1.49
2-5	2-left	9,000	0	12:58	13:14	16	7	0.43	0.19	0.29	0.25	21.75	0.09	0.05	18.55
3-1	2-heading	8,000	0	8:21	8:38	17	5	0.30	0.11	0.00	1.51	29.53	0.00	1.73	36.29
3-2	2-heading	7,900	6,500	9:44 11:10	10:00 11:14	20	9	0.44	0.02	0.00	0.82	2.21	0.00	0.67	1.84
3-3	1-left	7,700	0	11:47	12:09	22	9	0.41	0.23	0.46	4.64	30.34	0.20	3.98	27.17
3-4	3-heading	9,000	6,600	13:01	13:32	31	8	0.26	0.05	0.43	2.43	1.51	0.48	3.03	1.86
Avg	-	**8,350**	**6,430**	-	-	-	-	**0.37**	-	-	-	-	-	-	-

* Abbreviations: FBS, flooded-bed scrubber; CM, continuous miner; SC, shuttle car; RRC, right rear corner.

† Adjusted dust concentrations have intake levels subtracted and are normalized for differences in face airflow and production (SCs/min).

Table A-6. Shuttle car loading times and dust concentrations with the scrubber off

Cut No.	SC No.	SC begin loading	SC end loading	Intake dust, mg/m³	SC* dust, mg/m³
1-1	1	8:20:19	8:21:29	0.00	0.15
	2	8:22:24	8:23:19	0.00	0.11
	1	8:25:48	8:26:51	0.02	0.33
	1	8:29:03	8:30:13	0.00	0.45
	2	8:35:28	8:36:45	0.00	0.18
	1	8:37:16	8:38:55	0.00	1.16
	1	8:43:00	8:44:21	0.00	0.24
	1	8:46:24	8:47:23	0.00	0.32
	2	8:47:48	8:49:13	0.01	0.89
	1	8:50:20	8:51:49	0.00	1.16
	2	8:52:27	8:53:53	0.00	0.75
Cut avg	-	-	-	-	**0.52**
1-3	1	10:11:05	10:12:16	0.00	0.26
	2	10:13:00	10:14:06	0.00	0.53
	1	10:14:51	10:15:41	0.01	0.25
	2	10:17:16	10:18:10	0.00	0.81
	1	10:19:50	10:21:06	0.24	0.20
	2	10:21:54	10:22:46	0.01	0.52
	1	10:23:27	10:24:41	0.00	0.22
	2	10:26:05	10:27:16	0.00	0.36
	1	10:27:58	10:29:21	0.00	0.27
	2	10:30:35	10:31:11	0.00	0.30
Cut avg	-	-	-	-	**0.37**
1-5	2	12:18:05	12:19:38	0.00	0.13
	1	12:20:33	12:21:33	0.01	0.28
	2	12:22:40	12:23:30	0.01	0.24
	1	12:25:00	12:25:57	0.01	0.37
	2	12:27:14	12:28:48	0.00	0.29
	1	12:32:44	12:33:58	0.01	0.18
	2	12:36:18	12:37:58	0.00	0.10
	1	12:38:59	12:41:40	0.02	0.12
Cut avg	-	-	-	-	**0.21**
2-2	1	9:03:46	9:05:27	0.12	0.00
	2	9:05:57	9:07:29	0.07	0.02
	1	9:07:54	9:09:11	0.01	0.44
	2	9:10:44	9:11:54	0.00	0.73
	1	9:13:06	9:14:41	0.00	0.31
	2	9:15:54	9:17:13	0.02	0.11
	1	9:21:24	9:22:44	0.00	0.98
	2	9:23:01	9:25:39	0.01	0.24
	1	9:26:57	9:28:38	0.00	0.24
Cut avg	-	-	-	-	**0.34**

* Shuttle car (SC) dust levels shown have the intake dust subtracted; if less than zero, zero is shown.

28

Table A-6. Shuttle car loading times and dust concentrations with the scrubber off (Continued)

Cut No.	SC No.	SC begin loading	SC end loading	Intake dust, mg/m^3	SC* dust, mg/m^3
2-5	1	12:57:36	12:58:38	0.16	0.01
	2	12:59:13	12:59:59	0.24	0.59
	1	13:01:17	13:02:00	0.21	0.01
	2	13:03:33	13:04:35	0.18	0.05
	1	13:06:10	13:07:39	0.54	0.00
	2	13:10:18	13:11:24	0.07	0.07
	1	13:11:56	13:13:56	0.06	0.22
Cut avg	-	-	-	-	**0.14**
3-1	1	8:21:00	8:22:26	0.12	0.00
	2	8:24:09	8:25:01	0.12	0.03
	1	8:26:15	8:27:16	0.11	0.04
	2	8:28:44	8:31:06	0.11	0.05
	1	8:34:42	8:37:37	0.10	0.00
Cut avg	-	-	-	-	**0.02**
3-3	2	11:46:55	11:47:58	0.11	0.32
	1	11:48:58	11:49:49	0.16	0.00
	2	11:51:53	11:52:48	0.46	0.28
	1	11:53:45	11:54:39	0.59	0.14
	2	11:56:10	11:57:49	0.54	0.00
	1	11:59:17	12:00:30	0.16	0.94
	2	12:01:30	12:02:45	0.08	0.26
	1	12:03:48	12:05:16	0.05	0.36
	2	12:07:07	12:08:59	0.12	0.29
Cut avg	-	-	-	-	**0.29**
Survey avg	-	-	-	-	**0.27**

* Shuttle car (SC) dust levels shown have the intake dust subtracted; if less than zero, zero is shown.

Table A-7. Shuttle car loading times and dust concentrations with the scrubber on

Cut No.	SC No.	SC begin loading	SC end loading	Intake dust, mg/m^3	SC[*] dust, mg/m^3
1-2	1	9:20:01	9:21:42	0.05	0.09
	2	9:22:38	9:23:50	0.12	0.09
	1	9:25:08	9:26:40	0.06	0.48
	2	9:28:37	9:29:52	0.15	0.10
	1	9:31:43	9:33:53	0.21	0.00
	2	9:36:40	9:38:20	0.09	0.01
	1	9:40:08	9:41:17	0.00	0.41
	2	9:42:14	9:44:01	0.00	0.10
	1	9:45:05	9:46:32	0.00	0.11
Cut avg	-	-	-	-	**0.15**
1-4	1	11:14:13	11:15:28	0.00	0.07
	2	11:16:06	11:16:44	0.00	2.94
	1	11:18:17	11:19:23	0.00	0.29
	2	11:20:05	11:21:10	0.00	2.22
	1	11:24:55	11:25:55	0.00	0.13
	2	11:32:48	11:33:35	0.00	0.19
	1	11:34:18	11:35:22	0.00	0.28
	2	11:36:38	11:38:15	0.00	0.17
	1	11:40:25	11:42:15	0.00	0.10
	2	11:43:11	11:45:21	0.00	0.05
Cut avg	-	-	-	-	**0.64**
2-1	2	7:37:50	7:39:19	0.16	0.03
	2	7:47:38	7:48:28	0.21	0.66
	1	7:50:18	7:51:28	0.25	0.32
	2	7:53:06	7:54:16	0.11	0.00
	1	7:56:24	7:57:30	0.10	0.07
	2	7:58:52	7:59:52	0.09	0.06
	1	8:03:30	8:04:27	0.04	0.52
	2	8:08:27	8:09:49	0.06	0.39
	1	8:11:40	8:13:22	0.03	0.53
Cut avg	-	-	-	-	**0.29**
2-3	1	9:59:14	10:00:10	0.30	0.51
	2	10:01:16	10:02:00	0.39	0.16
	1	10:03:04	10:03:54	0.05	0.23
	2	10:06:30	10:07:20	0.01	0.20
	1	10:08:40	10:09:25	0.02	0.42
	2	10:12:44	10:13:50	0.09	0.00
	1	10:16:55	10:18:07	0.07	0.11
Cut avg	-	-	-	-	**0.23**

[*] Shuttle car (SC) dust levels shown have the intake dust subtracted; if less than zero, zero is shown.

Table A-7. Shuttle car loading times and dust concentrations with the scrubber on (Continued)

Cut No.	SC No.	SC begin loading	SC end loading	Intake dust, mg/m^3	SC* dust, mg/m^3
2-4	2	11:31:30	11:32:55	0.00	0.19
	1	11:33:57	11:35:32	0.00	0.51
	2	11:37:39	11:39:02	0.00	0.65
	1	11:39:41	11:40:59	0.00	0.64
	2	12:05:34	12:09:02	0.00	0.33
	1	12:09:50	12:11:18	0.00	0.64
	2	12:13:03	12:15:01	0.00	1.33
	1	12:18:20	12:19:52	0.00	0.20
	2	12:20:45	12:22:36	0.00	0.48
	1	12:23:25	12:25:13	0.00	0.28
Cut avg	-	-	-	-	**0.53**
3-2	2	9:43:46	9:44:32	0.03	0.02
	1	9:45:51	9:46:34	0.02	0.05
	2	9:48:29	9:49:27	0.02	0.03
	1	9:50:37	9:51:39	0.02	0.04
	2	9:54:15	9:55:07	0.02	0.02
	1	9:56:22	9:57:13	0.02	0.04
	2	9:59:32	10:00:24	0.02	0.10
	1	11:10:17	11:11:40	0.03	0.05
	2	11:12:56	11:14:14	0.02	0.44
Cut avg	-	-	-	-	**0.09**
3-4	1	13:00:45	13:01:41	0.11	0.00
	2	13:02:55	13:04:08	0.08	0.02
	1	13:04:51	13:06:13	0.05	0.00
	2	13:08:02	13:09:31	0.05	0.00
	1	13:10:08	13:12:45	0.05	0.02
	2	13:16:20	13:19:05	0.04	0.00
	1	13:19:45	13:21:34	0.04	0.02
	2	13:24:02	13:27:08	0.04	0.01
Cut avg	-	-	-	-	**0.01**
Survey avg	-	-	-	-	**0.28**

* Shuttle car (SC) dust levels shown have the intake dust subtracted; if less than zero, zero is shown.

Appendix B: Dust Survey at Mine B

Introduction

The purpose of this study was to compare respirable dust concentrations for two cutting conditions: (1) a standard 20-ft cut with a maximum 20-ft exhaust curtain setback and no flooded-bed scrubber operating, and (2) a 20-ft cut with an exhaust curtain setback up to 40 ft while operating a flooded-bed scrubber. A three-day dust survey was conducted alternating between both conditions on a cut-by-cut basis. This strategy was used to minimize the effects of changes in day-to-day operating conditions. For operating conditions known to affect face dust concentrations—including face airflow and the number, orientation, type, location, and pressure of external water sprays—attempts were made to maintain these conditions in a consistent state throughout the study.

Site Description

This dust survey was conducted on the right side of an eight-entry super section, as shown in Figure B-1. Coal was extracted using a Joy 12CM15 continuous mining machine and two Joy standard cab (off-curtain) shuttle cars. A dual-head, Fletcher DDR-13 roof bolter was used to install bolts. During the survey, entries 5 through 8 were developed, with intake air traveling up entry No. 5 and splitting to each side of the super section. Entries No. 8 and No. 9 (outby the faces being cut during the survey) were for return air. Mining height averaged 94 in, including the extraction of up to 30 in of top rock. The pillars were 60 ft x 80 ft with approximately 16-ft-wide entries. The continuous miner and roof bolter faces were ventilated with exhausting curtain hung on the right side of the entry.

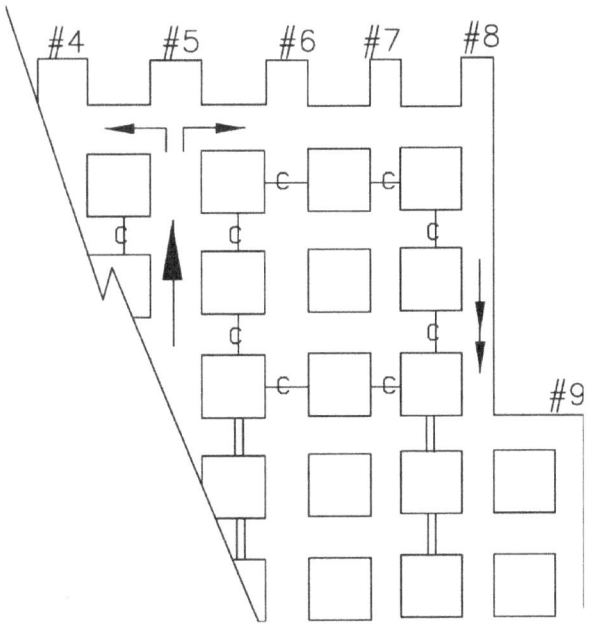

Figure B-1. Schematic of section sampled at Mine B.

Ventilation Plan Parameters

Prior to the NIOSH study, the mine had received approval from MSHA for a temporary ventilation plan that would be used throughout the study and allow for use of the scrubber. To assure consistent operating conditions, the mine's adherence to the parameters stipulated in the temporary ventilation plan was monitored on a periodic basis. Corrective actions were taken prior to initiating mining activity, when parameters fell outside required limits (less than the minimum specified or exceeding 120% of the minimum specified). The ventilation plan parameters required for this study included the following minimum ventilation quantities: 7,000 cfm at the face-side end of the brattice curtain on the continuous miner faces, 6,000 cfm scrubber airflow, and 5,000 cfm at the face-side end of the curtain on the bolter faces. A 30-layer pleated filter panel was used in the scrubber. Maximum curtain setbacks were 20 ft for the non-scrubber cuts and 40 ft for the scrubber cuts. The continuous miner was equipped with 42 water sprays (No. 3 hollow cone), with at least 36 sprays operational at all times. The minimum acceptable spray pressure was 65 psi.

Of the plan parameters monitored for this survey, obtaining airflow on the continuous miner faces and through the scrubber at the minimum or within 120% of the specified minimum level required the most effort. To meet plan parameters, face airflows between 7,000 and 8,400 cfm were required and were achieved in 12 of the 14 cuts sampled. The average curtain airflow during the survey was 7,943 cfm (SD = 390). Two cuts were out of the desired range, with airflow slightly exceeding the 8,400 cfm limit. Typically, an airflow reading was taken at the face in preparing for the upcoming cut and then adjustments to the ventilation curtains on the section had to be made in order to increase/decrease airflow to achieve the desired range.

The targeted scrubber airflow was between the 6,000 and 7,400 cfm. Measured scrubber airflows averaged 7,218 cfm (SD = 123), with only one reading falling above the targeted range by 24 cfm. However, it should be noted that in order to achieve suitable scrubber airflows it was necessary to clean the filter panel after each 20-ft cut. Once during each of the three days of sampling, the airflow capacity of the "dirty scrubber" was checked after the scrubber had been used to complete one 20-ft cut and prior to any cleaning. The airflow through the dirty scrubber had dropped an average of 2,090 cfm, which represented a 29% reduction from the initial airflow. The significant amount of rock cut in each face was thought to contribute to this drop in scrubber airflow.

The targeted curtain airflow on the bolting faces was between 5,000 and 6,000 cfm. For 11 of the 14 bolter faces, the measured airflow was within the desired range. For the other three places, airflow exceeded 6,000 cfm.

Water spray pressures were checked on the left and right side of the miner head twice during each shift. Operating pressure averaged 68 psi on both sides of the head. In general, mine personnel were able to adhere to the temporary ventilation plan approved for this dust survey.

Results

Use of real-time dust samplers enabled analysis of the data to be completed on a cut-by-cut basis. A total of 14 cuts were sampled, with the scrubber operated during seven of the cuts. Tables at the end of this appendix contain the raw data that were calculated for the various sampling instruments and sampling locations. Average dust concentrations were then calculated for scrubber-off and scrubber-on conditions in order to evaluate the impact on dust levels. Table B-1 summarizes cut locations and time, face and scrubber airflow, and dust concentrations generated around the continuous miner. The dual-numbering system for cuts represents the shift number followed by the number of the cut extracted during that shift. The miner operator dust concentrations were calculated from the PDM data, while data from all other sampling locations were calculated from pDR data. The dust concentrations for the miner operator, RRC, immediate return, and main return sampling locations had the intake dust levels subtracted out and were normalized for differences in productivity and face airflow, as discussed in the Data Analysis section of this report.

As shown in Table B-1c, all dust concentrations with the scrubber off were higher to varying degrees at the miner operator, rear corner, immediate return, and main return. However, only the 86% and 85% reductions in the return entries were statistically significant when using the Wilcoxon two-sample test, with $\alpha = 0.05$.

Table B-1. Summary of adjusted respirable dust concentrations for the continuous miner

a. Scrubber-off

Cut No.	Cut location	FBS* air-flow, cfm	Entry air-flow, cfm	Start cut	End cut	SCs*/min	CM* intake dust, mg/m³	CM oper dust, mg/m³	CM RRC* dust, mg/m³	CM return dust, mg/m³	Main return dust, mg/m³
1-1	8-heading	8,064	0	7:59	8:27	0.32	0.19	1.71	5.13	7.92	3.09
1-3	6-right	8,448	0	10:34	10:54	0.53	0.04	0.71	1.47	10.32	4.19
2-1	8-heading	8,640	0	7:57	8:29	0.37	0.18	0.21	23.22	10.37	5.82
2-3	7-heading	7,728	0	10:05	10:38	0.40	0.09	0.10	5.16	19.02	6.31
2-5	7-heading	8,400	0	12:38	13:02	0.42	0.08	0.18	14.62	16.36	6.50
3-1	6-right	7,105	0	7:44	8:15	0.39	0.17	0.09	1.21	12.91	2.67
3-3	7-right	7,680	0	11:08	11:46	0.34	0.12	0.88	1.88	8.04	5.08

*Abbreviations: FBS, flooded-bed scrubber; SC, shuttle car; CM, continuous miner; RRC, right rear corner.

b. Scrubber-on

Cut No.	Cut location	FBS air-flow, cfm	Entry air-flow, cfm	Start cut	End cut	SCs/min	CM intake dust, mg/m³	CM oper dust, mg/m³	CM RRC dust, mg/m³	CM return dust, mg/m³	Main return dust, mg/m³
1-2	7-heading	7,872	7424	9:28	9:55	0.33	0.17	0.98	6.82	2.21	0.78
1-4	6-right	7,920	7089	12:09	12:31	0.41	0.13	0.00	5.46	1.23	0.43
1-5	5-heading	8,204	7077	13:20	13:39	0.42	0.11	0.98	6.65	2.11	0.79
2-2	6-heading	7,808	7151	8:59	9:37	0.40	0.05	1.06	5.63	1.79	0.90
2-4	8-heading	7,840	7283	11:45	12:14	0.34	0.08	0.18	2.23	0.95	0.92
3-2	6-heading	7,813	7234	9:50	10:15	0.40	0.30	0.00	4.99	1.27	0.56
3-4	6-heading	7,673	7271	12:16	12:57	0.38	0.06	0.00	4.65	2.10	0.78

c. Impact of scrubber performance

Test condition and statistics	CM intake, mg/m^3	CM oper, mg/m^3	CM RRC, mg/m^3	CM return, mg/m^3	Main return, mg/m^3
Average dust concentration with scrubber off	0.12	0.55	7.53	12.13	4.81
Average dust concentration with scrubber on	0.13	0.46	5.20	1.67	0.74
Calculated p-value for cut data	-	*0.635*	*0.620*	*0.001*[†]	*0.001*[†]
Percent reduction in concentration with scrubber on	-	**16%**	**31%**	**86%**	**85%**

[†] Statistically significant difference.

Table B-2 summarizes the shuttle car dust concentrations collected during the study. These data isolate dust levels at the shuttle cars when located at the face during miner cutting and loading activities, and do not include exposures to dust while the cars are in transit or at the feeder/breaker. Intake dust concentrations were subtracted from shuttle car concentrations to arrive at adjusted shuttle car concentrations resulting from face activities. Dust concentrations in the shuttle car cabs were very low and averaged less than 0.1 mg/m^3, with one exception. Dust levels for cut 2-1 were substantially higher than for the other 13 cuts (shown in Table B-6) and were excluded from calculating the average dust level with the scrubber off. Cut 2-1 was the first lift in entry 8, with intake air coming through the last open crosscut. From the shuttle car data, it appears that a portion of the ventilating air was traveling down entry 8 and not flowing behind the return curtain. The dust levels at the RRC of the miner for this cut were also the highest levels observed during the survey, which supports the argument that dust-laden air from the face was traveling down the entry. The continuous miner operator was standing in the crosscut and therefore was not exposed to elevated dust rollback down the entry. Shuttle car dust concentrations when loading at the face averaged 0.07 mg/m^3 with the scrubber on versus 0.03 mg/m^3 with the scrubber off. This difference was not statistically significant using the Wilcoxon test (*p-value = 0.4272*).

Table B-2. Summary of adjusted respirable dust concentrations in the shuttle car cabs

Cut number with scrubber off	Cut location with scrubber off	SC[*] dust with scrubber off, mg/m^3	Cut number with scrubber on	Cut location with scrubber on	SC dust with scrubber on, mg/m^3
1-1	8-heading	0.05	1-2	7-heading	0.04
1-3	6-right	0.04	1-4	6-right	0.01
2-1	8-heading	na	1-5	5-heading	0.17
2-3	7-heading	0.02	2-2	6-heading	0.06
2-5	7-heading	0.06	2-4	8-heading	0.18
3-1	6-right	0.02	3-2	6-heading	0.02
3-3	7-right	0.01	3-4	6-heading	0.04
Average	-	**0.03**	-	-	**0.07**

[*] Abbreviation: SC, shuttle car.

Table B-3 shows respirable dust concentration in the intake air supplying the roof bolter places and at the mid-bolter in the vicinity of the roof bolter operators. The intake concentrations were subtracted from the mid-bolter concentrations to estimate bolter-generated dust. The bolting machine generated very little dust (0.20 mg/m^3) with proper face ventilation maintained and the dust collector equipment in good working order. The dust collector vacuum pressures averaged 14 in Hg (SD = 0.7) on the intake-side of the bolter and 17 in Hg (SD = 0.5) on the return-side. The minimum required pressure was 12 in Hg.

Table B-3. Summary of roof bolter dust concentrations

a. Bolter operating location and dust levels

Place No.	Face location	Face airflow, cfm	Position of RB* with respect to CM*	Start time	End time	Scrubber status	RB intake dust, mg/m^3	RB center dust, mg/m^3
1-1	7-heading	5,515	upwind	7:30	8:05	na	0.12	0.42
1-2	8-heading	5,550	miner off	8:56	9:28	na	0.06	0.29
1-3	7-heading	6,000	downwind	10:26	10:52	off	4.75	3.18
1-4	6-right	7,480	miner off	11:26	11:54	na	0.06	0.31
1-5	6-right	5,880	miner off	12:50	13:23	na	0.14	0.54
2-1	7-heading	5,376	upwind	7:31	8:06	na	0.04	0.02
2-2	8-heading	5,670	combined†	9:30	10:05	on	0.76	0.90
2-3	6-heading	5,760	upwind	10:28	11:00	na	0.19	0.39
2-4	7-heading	6,000	upwind	11:16	11:51	na	0.18	0.16
2-5	8-heading	5,880	downwind	12:36	13:07	off	7.18	8.12
3-1	6-right‡	5,670	miner off	8:35	8:49	na	0.16	0.25
3-2	6-right	21,560	downwind	9:43	10:19	on	1.01	0.85
3-3	6-heading	5,712	upwind	10:55	11:29	na	0.09	0.28
3-4	7-right	19,296	downwind	12:05	12:54	on	1.00	1.06

* Abbreviations: RB, roof bolter; CM, continuous miner.

† Continuous miner operating upwind of bolter during first 7 min of bolter operation.

‡ Partial cut—only 3 rows of bolts installed.

b. Impact of continuous miner on dust levels at the roof bolter

RB position and status of CM	RB intake, mg/m^3	RB center, mg/m^3
Average dust concentration when RB upwind or with CM not operating	0.12	0.30
Average dust concentration when RB downwind of CM with scrubber off	5.97	5.65
Average dust concentration when RB downwind of CM with scrubber on	0.92	0.94
Percent dust reduction when downwind of CM with scrubber on	**85%**	**83%**

Past research has shown that significant exposures to respirable dust can occur when the bolting machine operates downwind of the continuous mining machine [Jayaraman et al. 1988]. During this survey, the bolter was downwind of the miner during five different cuts. For two of these cuts, the scrubber was not operating and dust levels measured by the intake bolter package averaged 5.97 mg/m^3. The scrubber was operating during three cuts with the bolter downwind and average intake dust levels were 0.92 mg/m^3. Although an 85% difference in dust levels was observed, the small number of samples did not warrant testing for a statistically significant difference. A similar difference was observed at the sampling location on the bolter.

Table B-4 displays the mass, percent, and concentration of quartz for miner-generated respirable dust from the immediate return for samples collected for both the scrubber-off and scrubber-on conditions. Despite a substantial amount of rock being cut, the quartz levels were less than 4%. However, use of the scrubber did reduce the average miner-generated quartz dust mass by 83% from 38 μg to 7 μg. In order to account for differences in productivity, the quartz concentrations were normalized relative to the shuttle cars/min for each cut. As shown in Table B-4c, the adjusted quartz concentrations illustrated a difference of 82%, which was statistically significant (Wilcoxon test).

Table B-4. Respirable quartz levels in the continuous miner return

a. Scrubber-off

Shift	Gravimetric filter number	Sampling time, min	Dust mass, mg	Avg dust conc, mg/m^3	Quartz mass, μg	Quartz, %	Quartz conc, μg/m^3	Shuttle cars/ min	Adjusted quartz conc*, μg/m^3
1	629561	71	0.786	5.54	30	3.8	211.27	0.40	200.70
	629628	71	0.859	6.05	30	3.5	211.27	0.40	200.70
2	629566	67	1.260	9.40	39	3.1	291.04	0.39	283.58
	629573	67	1.270	9.48	49	3.9	365.67	0.39	356.30
3	566996	87	1.091	6.27	41	3.8	235.63	0.36	248.72
	629565	46	0.555	6.03	21	3.8	228.26	0.36	240.94

* Concentrations adjusted based upon an average of 0.38 shuttle cars per min.

b. Scrubber-on

Shift	Gravimetric filter number	Sampling time, min	Dust mass, mg	Avg dust conc, mg/m^3	Quartz mass, μg	Quartz, %	Quartz conc, μg/m^3	Shuttle cars/ min	Adjusted quartz conc*, μg/m^3
1	629570	81	0.277	1.71	8	2.9	49.38	0.38	49.38
	629617	81	0.288	1.78	11	3.8	67.90	0.38	67.90
2	629557	78	0.240	1.54	8	3.3	51.28	0.37	52.67
	629599	78	0.229	1.47	7	3.1	44.87	0.37	46.08
3	566992	76	0.209	1.38	3	1.4	19.74	0.39	19.23
	566987	76	0.217	1.43	5	2.3	32.89	0.39	32.05

* Concentrations adjusted based upon an average of 0.38 shuttle cars per min.

c. Impact of scrubber performance

Test condition and statistics	Quartz mass, μg	Quartz conc, μg/m^3	Adjusted quartz conc, μg/m^3
Average quartz content with scrubber off	38	257.19	255.16
Average quartz content with scrubber on	7	44.34	44.55
Calculated p-value	-	-	*0.0022†*
Percent reduction in quartz content with scrubber on	**82%**	**83%**	**82%**

† Statistically significant difference.

Conclusions

The purpose of the study was to compare face dust levels for two cutting conditions: (1) a standard 20-ft cut with a maximum 20-ft curtain setback and no scrubber operating, and (2) a 20-ft advance with curtain setbacks of up to 40 ft while operating a flooded-bed scrubber. Because the approved ventilation and dust control plan for this mechanized mining unit (MMU) did not contain provisions to use a flooded-bed scrubber, the mine obtained a temporary plan for use during NIOSH testing. To assure consistent operating conditions throughout the survey, MSHA personnel assisted in the study by monitoring the mine's adherence to the dust control parameters stipulated in the temporary ventilation plan. Airflow adjustments were made by mine personnel throughout the survey in an effort to maintain levels within required limits (minimum specified in the plan or up to 120% of the minimum specified). In general, mine personnel were successful in meeting the conditions specified in the temporary ventilation plan. Key findings from the survey include:

- Average respirable dust concentrations were 86% lower behind the miner return curtain with the scrubber operating, decreasing from 12.13 mg/m^3 to 1.66 mg/m^3. This difference is statistically significant (Wilcoxon test, $\alpha = 0.05$).

- Average respirable dust concentrations were 85% lower in the main return with the scrubber operating, decreasing from 4.81 mg/m^3 to 0.74 mg/m^3. This difference is statistically significant (Wilcoxon test).

- Average quartz concentration in the immediate miner return was reduced by 82% with the scrubber operating, decreasing from 255 µg/m^3 to 45 µg/m^3. This difference is statistically significant (Wilcoxon test).

- Average dust concentrations at the continous miner operator (0.55 vs. 0.46 mg/m^3) and at the right rear corner of the mining machine (7.53 vs. 5.20 mg/m^3) were lower with the scrubber operating, but the differences are not statiscally significant (Wilcoxon test).

- Average dust concentrations in the shuttle car cabs when loading at the face averaged 0.07 mg/m^3 with the scrubber on versus 0.03 mg/m^3 with the scrubber off, but this difference is not statistically significant (Wilcoxon test).

- Dust concentrations generated by the bolting machine averaged 0.20 mg/m^3.

- The bolter operated downwind of the miner for five cuts during the dust survey. For two of these cuts, the scrubber was not operated and dust concentrations at the bolter intake averaged 5.97 mg/m^3. For the other three cuts, the scrubber was operated and dust concentrations at the bolter intake averaged 0.92 mg/m^3. This 85% reduction was not tested for statistical significance but was consistent with the statistically significant differences measured in the returns when comparing the scrubber-off and scrubber-on conditions.

The data collected for this study indicate that operation of the flooded-bed scrubber with an extended curtain setback did not result in a statistically significant difference in dust concentrations at the continuous miner operator and shuttle car operator positions when compared to dust concentrations generated with standard cutting conditions. However, operation of the scrubber did result in at least an 85% reduction in respirable dust in the return airstreams and an 82% reduction in respirable quartz. These differences were statically significant. It should be noted that these results were obtained with the MMU operating under conditions specified in a temporary ventilation plan that was approved for use during this survey.

The statistically significant differences measured in respirable dust concentrations downwind from the miner when the scrubber was operating support a recommendation that mine management seek to use the flooded-bed scrubber on a regular basis. Reduced respirable and quartz dust exposures could be expected for mine personnel positioned downwind of the continuous miner. Although not statistically significant, dust concentrations around the miner were also lower with the scrubber operating.

Table B-5. Continuous-miner-generated dust concentrations (mg/m³) for each cut

Cut No.	Cut location	Face air-flow, cfm	FBS* air-flow, cfm	CM* start time	CM stop time	Total cut time, min	No. SC*	SCs/ min	CM int dust conc	CM oper dust conc	CM RRC* dust conc	CM ret dust conc	Main ret dust conc	Adj† CM oper dust conc	Adj† CM RRC dust conc	Adj† CM ret dust conc	Adj† main ret dust conc
1-1	8-heading	8,064	0	7:59	8:27	28	9	0.32	0.19	1.58	4.36	6.62	2.70	1.71	5.13	7.92	3.09
1-2	7-heading	7,872	7,424	9:28	9:55	27	9	0.33	0.17	1.02	6.06	2.08	0.85	0.98	6.82	2.21	0.78
1-3	6-right	8,448	0	10:34 10:46	10:43 10:54	17	9	0.53	0.04	0.95	1.91	13.23	5.39	0.71	1.47	10.32	4.19
1-4	6-right	7,920	7,089	12:09	12:31	22	9	0.41	0.13	0.13	5.88	1.42	0.58	0.00	5.46	1.23	0.43
1-5	5-heading	8,204	7,077	13:20	13:39	19	8	0.42	0.11	1.14	7.07	2.32	0.94	0.98	6.65	2.11	0.79
2-1	8-heading	8,640	0	7:57 8:15	8:10 8:29	27	10	0.37	0.18	0.37	20.47	9.24	5.27	0.21	23.22	10.37	5.82
2-2	6-heading	7,808	7,151	8:59 9:06 9:12 9:20	9:03 9:08 9:14 9:37	25	10	0.40	0.05	1.17	5.94	1.93	1.00	1.06	5.63	1.79	0.90
2-3	7-heading	7,728	0	10:05 10:18	10:10 10:38	25	10	0.40	0.09	0.20	5.54	20.16	6.75	0.10	5.16	19.02	6.31
2-4	8-heading	7,840	7,283	11:45	12:14	29	10	0.34	0.08	0.24	2.08	0.93	0.91	0.18	2.23	0.95	0.92
2-5	7-heading	8,400	0	12:38	13:02	24	10	0.42	0.08	0.26	14.86	16.61	6.65	0.18	14.62	16.36	6.50
3-1	6-right	7,105	0	7:44 7:53 8:03	7:49 7:59 8:15	23	9	0.39	0.17	0.26	1.52	14.66	3.17	0.09	1.21	12.91	2.67
3-2	6-heading	7,813	7,234	9:50	10:15	25	10	0.40	0.30	0.00	5.50	1.62	0.88	0.00	4.99	1.27	0.56
3-3	7-right	7,680	0	11:08 11:18	11:15 11:46	35	12	0.34	0.12	0.92	1.83	7.44	4.74	0.88	1.88	8.04	5.08
3-4	6-heading	7,673	7,271	12:16 12:36	12:19 12:57	24	9	0.38	0.06	0.00	4.69	2.15	0.84	0.00	4.65	2.10	0.78
Avg	-	**7,943**	**7,218**	-	-	-	-	**0.39**	-	-	-	-	-	-	-	-	-

* Abbreviations: FBS, flooded-bed scrubber; CM, continuous miner; SC, shuttle car; RRC, right rear corner.
† Adjusted dust concentrations have intake levels subtracted and are normalized for differences in face airflow and production (SCs/min).

41

Table B-6. Shuttle car loading times and dust concentrations with the scrubber off

Cut No.	SC No.	SC begin loading	SC end loading	Intake dust, mg/m^3	SC[*] dust, mg/m^3
1-1	1	7:59:03	7:59:58	0.19	0.00
	2	8:01:52	8:02:45	0.17	0.00
	1	8:05:36	8:06:12	0.20	0.00
	2	8:08:53	8:09:48	0.14	0.00
	1	8:11:52	8:12:58	0.08	0.00
	2	8:14:50	8:15:57	0.48	0.00
	1	8:17:55	8:18:45	0.34	0.25
	2	8:20:42	8:21:40	0.20	0.03
	1	8:23:43	8:24:46	0.11	0.18
	1	8:26:07	8:26:40	0.09	0.00
Cut avg	-	-	-	-	**0.05**
1-3	2	10:34:00	10:35:10	0.00	0.02
	1	10:36:17	10:37:10	0.06	0.12
	2	10:38:08	10:39:00	0.03	0.05
	1	10:39:50	10:40:45	0.06	0.04
	2	10:41:32	10:42:16	0.02	0.07
	1	10:46:23	10:47:37	0.01	0.01
	2	10:48:30	10:49:42	0.02	0.04
	1	10:50:38	10:51:37	0.05	0.04
	2	10:52:32	10:53:21	0.02	0.03
Cut avg	-	-	-	-	**0.05**
2-1	1	7:57:11	7:58:21	0.15	0.60
	2	7:59:52	8:00:35	0.13	6.91
	1	8:01:52	8:02:36	0.19	2.76
	2	8:03:58	8:05:03	0.16	1.08
	1	8:06:23	8:07:31	0.23	1.17
	2	8:08:54	8:09:43	0.18	0.85
	1	8:15:00	8:16:43	0.42	5.10
	2	8:18:16	8:19:37	0.23	17.84
	1	8:21:07	8:22:15	0.14	2.79
	2	8:23:38	8:24:25	0.12	2.57
	2	8:26:11	8:26:44	0.11	0.36
	2	8:28:00	8:28:37	0.11	0.20
Cut avg	-	-	-	-	**3.52[†]**
2-3	1	10:05:50	10:07:04	0.00	0.03
	2	10:08:06	10:09:01	0.02	0.02
	1	10:18:25	10:19:34	0.06	0.06
	2	10:20:35	10:21:35	0.11	0.00
	1	10:22:45	10:23:40	0.05	0.01
	2	10:24:29	10:25:17	0.04	0.05
	1	10:26:26	10:27:12	0.07	0.00
	2	10:28:57	10:30:40	0.07	0.00
	1	10:32:05	10:33:35	0.18	0.00

[*] Shuttle car (SC) dust levels have intake levels subtracted; if less than zero, zero is shown.

[†] Data excluded from calculation of average dust concentration for cuts with the scrubber off.

Table B-6. Shuttle car loading times and dust concentrations with the scrubber off (Continued)

Cut No.	SC No.	SC begin loading	SC end loading	Intake dust, mg/m^3	SC* dust, mg/m^3
2-3	2	10:34:58	10:35:45	0.26	0.00
	2	10:36:49	10:37:37	0.30	0.00
Cut avg	-	-	-	-	**0.02**
2-5	1	12:38:17	12:39:27	0.07	0.00
	2	12:40:18	12:40:47	0.01	0.05
	1	12:41:59	12:43:02	0.03	0.46
	2	12:45:28	12:46:51	0.19	0.00
	1	12:49:14	12:49:57	0.03	0.00
	2	12:51:00	12:51:49	0.02	0.02
	1	12:53:10	12:53:43	0.02	0.07
	2	12:55:05	12:56:25	0.08	0.00
	1	12:57:22	12:58:24	0.15	0.00
	2	12:59:33	13:00:00	0.03	0.02
	2	13:01:29	13:01:50	0.28	0.00
Cut avg	-	-	-	-	**0.06**
3-1	2	7:44:14	7:45:13	0.14	0.00
	1	7:48:01	7:48:44	0.14	0.02
	2	7:53:04	7:54:04	0.10	0.00
	1	7:55:32	7:56:22	0.13	0.11
	2	7:58:10	7:59:00	0.13	0.00
	1	8:03:11	8:04:20	0.12	0.02
	2	8:05:48	8:06:58	0.14	0.00
	1	8:09:27	8:10:33	0.27	0.02
	2	8:12:03	8:13:15	0.23	0.00
	2	8:14:17	8:15:01	0.17	0.00
Cut avg	-	-	-	-	**0.02**
3-3	2	11:08:14	11:09:23	0.12	0.00
	1	11:11:37	11:12:38	0.16	0.00
	2	11:14:25	11:15:00	0.16	0.00
	1	11:18:02	11:18:45	0.21	0.00
	2	11:20:28	11:21:21	0.17	0.00
	1	11:23:37	11:24:25	0.14	0.00
	2	11:27:12	11:28:03	0.16	0.00
	1	11:30:34	11:32:03	0.11	0.02
	2	11:33:36	11:34:25	0.09	0.00
	1	11:36:55	11:37:42	0.07	0.05
	2	11:39:31	11:40:27	0.06	0.00
	1	11:43:30	11:44:12	0.07	0.00
	1	11:45:07	11:45:35	0.06	0.00
Cut avg	-	-	-	-	**0.01**
Survey avg	-	-	-	-	**0.04**

* Shuttle car (SC) dust levels have intake levels subtracted; if less than zero, zero is shown.
† Data excluded from calculation of average dust concentration for cuts with the scrubber off.

Table B-7. Shuttle car loading times and dust concentrations with the scrubber on

Cut No.	SC No.	SC begin loading	SC end loading	Intake dust, mg/m^3	SC* dust, mg/m^3
1-2	1	9:28:50	9:29:35	0.06	0.05
	2	9:31:20	9:32:00	0.33	0.00
	1	9:35:10	9:35:52	0.29	0.00
	2	9:37:40	9:38:25	0.18	0.01
	1	9:40:05	9:40:58	0.12	0.17
	2	9:43:18	9:44:40	0.04	0.03
	1	9:46:27	9:47:15	0.09	0.05
	2	9:50:00	9:51:00	0.24	0.00
	1	9:53:24	9:54:07	0.03	0.08
Cut avg	-	-	-	-	**0.04**
1-4	2	12:09:08	12:10:10	0.05	0.00
	1	12:11:08	12:11:49	0.49	0.00
	2	12:12:45	12:13:40	0.23	0.00
	1	12:15:08	12:15:57	0.18	0.01
	2	12:18:23	12:19:30	0.21	0.00
	1	12:20:32	12:21:49	0.07	0.00
	2	12:22:52	12:24:25	0.10	0.00
	1	12:25:44	12:26:38	0.02	0.03
	2	12:28:13	12:29:00	0.02	0.03
	2	12:30:10	12:30:47	0.03	0.00
Cut avg	-	-	-	-	**0.01**
1-5	1	13:20:50	13:21:57	0.02	0.02
	2	13:24:09	13:24:59	0.10	0.05
	1	13:25:38	13:26:26	0.26	0.18
	2	13:27:13	13:28:21	0.03	0.36
	1	13:29:30	13:30:25	0.07	0.04
	2	13:31:14	13:32:10	0.02	0.38
	1	13:32:52	13:33:45	0.21	0.41
	2	13:36:00	13:36:51	0.09	0.04
	2	13:37:57	13:38:43	0.04	0.03
Cut avg	-	-	-	-	**0.17**
2-2	2	8:59:48	9:00:47	0.07	0.13
	1	9:02:10	9:02:47	0.07	0.26
	2	9:06:39	9:07:33	0.07	0.03
	1	9:12:11	9:13:05	0.07	0.00
	2	9:20:23	9:21:17	0.07	0.04
	1	9:23:13	9:24:21	0.07	0.00
	2	9:25:44	9:26:49	0.07	0.00
	1	9:28:05	9:29:20	0.07	0.00
	2	9:31:05	9:32:23	0.07	0.26
	1	9:34:03	9:34:55	0.07	0.00
	1	9:36:03	9:36:50	0.07	0.00
Cut avg	-	-	-	-	**0.07**

* Shuttle car (SC) dust levels have intake levels subtracted; if less than zero, zero is shown.

Table B-7. Shuttle car loading times and dust concentrations with the scrubber on (Continued)

Cut No.	SC No.	SC begin loading	SC end loading	Intake dust, mg/m^3	SC[*] dust, mg/m^3
2-4	1	11:45:00	11:45:50	0.07	0.08
	1	11:47:29	11:48:21	0.07	0.00
	2	11:49:57	11:50:36	0.07	0.00
	1	11:53:29	11:54:15	0.07	0.00
	1	11:55:55	11:56:36	0.07	0.31
	2	11:58:45	11:59:56	0.07	0.32
	1	12:01:54	12:03:10	0.07	0.26
	2	12:05:19	12:06:22	0.07	0.59
	1	12:08:19	12:09:17	0.07	0.13
	2	12:11:06	12:11:45	0.07	0.05
	2	12:13:07	12:13:47	0.07	0.00
Cut avg	-	-	-	-	**0.16**
3-2	1	9:50:30	9:51:32	0.73	0.00
	2	9:52:28	9:53:14	0.25	0.00
	1	9:55:00	9:55:50	0.27	0.00
	2	9:56:33	9:57:42	0.25	0.00
	1	9:59:09	10:00:10	0.96	0.00
	2	10:00:59	10:01:59	0.26	0.00
	1	10:03:35	10:04:30	0.16	0.00
	2	10:05:23	10:06:17	0.09	0.01
	1	10:07:53	10:08:53	0.06	0.16
	2	10:10:56	10:11:53	0.04	0.00
	2	10:13:38	10:14:23	0.06	0.00
Cut avg	-	-	-	-	**0.02**
3-4	1	12:16:23	12:17:06	0.04	0.03
	2	12:18:01	12:19:01	0.06	0.00
	1	12:36:23	12:37:08	0.02	0.07
	2	12:38:20	12:39:20	0.05	0.05
	1	12:40:35	12:41:50	0.04	0.00
	2	12:43:02	12:43:52	0.05	0.02
	1	12:47:17	12:48:14	0.03	0.07
	2	12:50:20	12:51:04	0.08	0.00
	1	12:53:58	12:54:43	0.02	0.18
	1	12:55:44	12:56:30	0.22	0.00
Cut avg	-	-	-	-	**0.04**
Survey avg	-	-	-	-	**0.07**

[*] Shuttle car (SC) dust levels have intake levels subtracted; if less than zero, zero is shown.

Appendix C: Dust Survey at Mine C

Introduction

The purpose of this study was to compare respirable dust levels from a standard 20-ft cut with a maximum 20-ft exhaust curtain setback and no flooded-bed scrubber operating to a 20-ft cut with an exhaust curtain setback of 40 ft while operating a flooded-bed scrubber. A three-day dust survey was conducted, alternating between both test conditions on a cut-by-cut basis. This strategy was designed to minimize the effects of changes in day-to-day operating conditions. Unfortunately, a number of the cuts had atypical ventilation patterns in the face, which complicated comparison of the data. For operating conditions known to affect face dust concentrations—including face airflow and the number, orientation, type, location, and pressure of external water sprays—attempts were made to maintain these conditions in a consistent state throughout the study.

Site Description

This dust survey was conducted on the left side of a nine-entry super section as shown in Figure C-1. Coal was extracted using a Joy 14 CM 15 continuous mining machine and two Joy standard cab (off-curtain) shuttle cars. A dual-head, Fletcher Roof Ranger II roof bolter was used to install bolts. During the survey, the miner developed entries 1 through 5, with intake air traveling up entries 3 and 4 and entries 1 and 2 used for return air. Entry 5 contained the feeder/breaker and section belt and was isolated in the ventilation plan. Mining height averaged 65 inches, including the extraction of up to three ft of rock. The miner was cutting 20-ft-wide entries. The faces were ventilated with exhausting curtain hung on the left side of the entry.

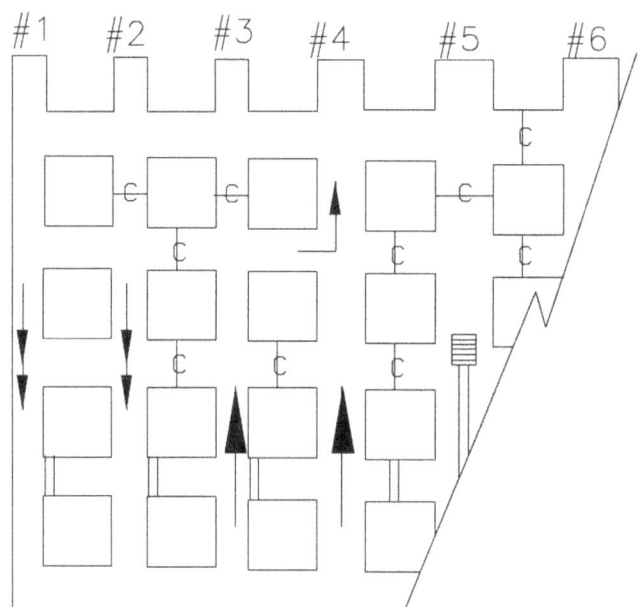

Figure C-1. Schematic of section sampled at Mine C.

46

Challenging mining conditions were encountered during the survey with a steep rise in the coal seam, which made shuttle car travel and cutting with the continuous miner more difficult. As a result, the mining unit did not achieve normal levels of production during the NIOSH sampling period.

Ventilation Plan Parameters

In an effort to assure consistent operating conditions, the mine's adherence to the parameters stipulated in the ventilation plan was monitored by MSHA personnel. Corrective actions were taken prior to initiating mining activity when parameters fell outside required limits (less than the minimum specified in the plan or greater than 120% of plan parameters). The ventilation plan parameters required on this MMU included the following minimum ventilation quantities: 5,000 cfm at the face-side end of the brattice curtain on the continuous miner faces, 4,000 cfm scrubber airflow, and 4,000 cfm at the face-side end of the curtain on the bolter faces. A 30-layer pleated filter panel was used in the scrubber. Maximum curtain setbacks were 20 ft for the scrubber-off cuts and 40 ft for the scrubber-on cuts. The continuous miner was equipped with 33 water sprays (BD 3-5 hollow cone), with at least 30 sprays operational at all times. The minimum allowable spray pressure was 85 psi.

For the plan parameters monitored during the survey, obtaining the required airflow on the continuous miner faces was the most challenging. Typically, an airflow reading was taken at the face in preparing for the upcoming cut and then adjustments to the ventilation curtains on the section had to be made in order to increase/decrease airflow to achieve the desired range. To meet plan parameters, face airflows between 5,000 and 6,000 cfm were desired and were achieved in 4 of the 8 cuts sampled. The average curtain airflow during the survey was 6,565 cfm (SD = 1396). Four cuts were out of the desired range, with measured airflow exceeding the 6,000 cfm upper limit.

The targeted scrubber airflow was between 4,000 and 4,800 cfm. Measured scrubber airflows averaged 4,288 (SD = 131), with all readings falling within the targeted range. However, it should be noted that in order to achieve suitable scrubber airflows it was necessary to clean the 30-layer filter panel after each 20-ft cut. Once during the survey, the airflow capacity of the "dirty scrubber" was checked after the scrubber had been used to complete one 20-ft cut and prior to any cleaning. The airflow through the dirty scrubber had dropped 1,549 cfm, which represented a 35% reduction from the initial airflow. The significant amount of rock being cut in each face was thought to contribute to this substantial drop in scrubber airflow.

The minimum required curtain airflow on the bolting faces was 4,000 cfm. For all seven of the bolter faces, the measured airflow exceeded this quantity.

Water spray pressures on the miner were checked during each shift to ensure the 85 psi minimum pressure was achieved. Operating pressure averaged 101 psi and 99 psi on the left and right sides of the miner, respectively. In addition, the required number of sprays was operating during each cut.

Results

Utilization of real-time samplers allowed for data analysis on a cut-by-cut basis. A total of eight cuts were sampled, with the scrubber operated during four of the cuts. The raw data that were calculated for the various sampling instruments and sampling locations are included at the end of this appendix. These raw data were used to calculate average dust concentrations for scrubber-off and scrubber-on test conditions in order to evaluate the scrubber impact on dust levels. Table C-1 summarizes cut locations and time, face and scrubber airflow, and dust concentrations generated around the continuous miner. The miner operator dust concentrations were calculated from the PDM data, while data from all other sampling locations were calculated from pDR data. Since the continuous miner was mining uphill, water from the boom sprays was draining down the top deck of the miner and onto the sampling package located at the right rear corner of the machine. This water was running onto the pDR sampler, which caused the pDR to stop operating during each of the first two sampling shifts. Therefore, it was decided to abandon this sampling location for shift 3, so no valid data were obtained at the RRC of the miner. The dust concentrations for the miner operator, immediate miner return, and main return sampling locations had the intake dust concentrations subtracted out and then were normalized for differences in productivity and for differences in face airflow, assuming a linear relationship between these parameters and dust levels. For example, average productivity for all cuts sampled during the survey was 0.28 shuttle cars per min (Table C-5), while productivity for Cut 1-1 averaged 0.36 shuttle cars per min. Dust levels for Cut 1-1 were then multiplied by a factor of 0.78 ($0.28 \div 0.36$) to adjust for the higher-than-average productivity observed in Cut 1-1. The airflow was adjusted similarly, with the factor calculated by dividing the measured face airflow by the average face airflow of all cuts (Table C-5).

Table C-1. Summary of adjusted respirable dust concentrations for the continuous miner

a. Scrubber-off

Cut No.	Cut location	FBS[*] air-flow, cfm	Entry air-flow, cfm	Start cut	End cut	SCs[*]/ min	CM[*] intake dust, mg/m³	CM oper dust, mg/m³	CM RRC[*] dust[†], mg/m³	CM return dust, mg/m³	Main return dust, mg/m³
1-2	5-heading	0	5,070	8:54	10:41	0.24	0.06	0.23	na	21.16	6.67
2-1	2-heading	0	6,348	7:47	8:56	0.32	0.02	0.02	na	5.63	4.58
2-2	1-heading	0	5,280	9:15	10:42	0.29	0.26	0.61	na	5.47	8.26
3-2	3-left	0	9,200	11:20	12:56	0.35	0.03	0.12	na	25.21	7.40

[*]Abbreviations: FBS, flooded-bed scrubber; SC, shuttle car; CM, continuous miner; RRC, right rear corner.
[†] Water inundation of samplers at this location caused data to be lost.

b. Scrubber-on

Cut No.	Cut location	FBS air-flow, cfm	Entry air-flow, cfm	Start cut	End cut	SCs/ min	CM intake dust, mg/m³	CM oper dust, mg/m³	CM RRC dust[†], mg/m³	CM return dust, mg/m³	Main return dust, mg/m³
1-1	3-heading	4,300	5,757	7:58	8:22	0.36	0.02	0.49	na	9.55	na
1-3	4-heading	4,400	5,910	11:34	12:19	0.33	0.05	na	na	7.17	3.29
2-3	3-heading	4,360	7,560	11:52	12:58	0.23	0.10	na	na	9.04	8.19
3-1	5-heading	4,100	7,398	9:18	10:16	0.15	0.03	0.14	na	8.81	2.91

[†] Water inundation of samplers at this location caused data to be lost.

c. Impact of scrubber performance

Test condition and statistics	CM intake, mg/m^3	CM oper, mg/m^3	CM RRC[†], mg/m^3	CM return, mg/m^3	Main return, mg/m^3
Average dust concentration with scrubber off	0.09	0.25	na	14.37	6.73
Average dust concentration with scrubber on	0.05	0.32	na	8.64	4.80
Calculated p-value for cut data	-	*0.8000*	na	*1.0000*	*0.4000*
Percent reduction in concentration with scrubber on	-	**-28%**	na	**40%**	**29%**

[†] Water inundation of samplers at this location caused data to be lost.

As shown in Table C-1, the dust concentrations around the continuous miner and in the main return exhibited substantial variability within each test condition. Given this variability and the limited amount of data, no statistically significant differences (Wilcoxon test with $\alpha = 0.05$) were observed between the scrubber-off and scrubber-on test conditions at the miner operator, the miner return, or the main return. Results show a 0.07-mg/m^3 increase in average dust levels at the miner operator when comparing scrubber-on to scrubber-off, while average dust levels in the immediate and main returns show reductions of over 3 mg/m^3 and nearly 2 mg/m^3, respectively.

Table C-2 summarizes the shuttle car dust concentrations collected during the study. These data isolate the dust levels at the shuttle cars when the cars are located at the face during miner cutting and loading activities, but do not include exposures while in transit or at the feeder/breaker. Intake dust concentrations were subtracted from shuttle car concentrations to arrive at adjusted shuttle car concentrations resulting from face activities. Dust concentrations in the shuttle car cabs also exhibited a substantial amount of variability within each test condition. Once again, no statistically significant differences were found between the test conditions with the Wilcoxon test (*p-value =0.5429*), with only a 0.08-mg/m^3 difference in average dust concentrations measured in the shuttle car cabs.

Table C-2. Summary of adjusted respirable dust concentrations in the shuttle car cabs

Cut number with scrubber off	Cut location with scrubber off	SC[*] dust with scrubber off, mg/m^3	Cut number with scrubber on	Cut location with scrubber on	SC dust with scrubber on, mg/m^3
1-2	5-heading	1.29	1-1	3-heading	0.23
2-1	2-heading	0.02	1-3	4-heading	1.33
2-2	1-heading	0.01	2-3	3-heading	0.01
3-2	3-left	0.07	3-1	5-heading	0.15
Average	-	**0.35**	-	-	**0.43**

[*] Abbreviation: SC, shuttle car.

Table C-3 shows respirable dust concentrations in the intake air supplying the roof bolter places and in the bolter cab. These dust concentrations were normalized for differences in face airflow similarly to the miner dust levels discussed previously. The dust collector vacuum pressures averaged 15 in Hg (SD = 0.7) on the left side of the bolter and 13 in Hg (SD = 0.5) on the right side. The minimum required pressure was 12 in Hg.

Past research has shown that significant exposures to respirable dust can occur when the bolting machine operates downwind of the continuous mining machine [Jayaraman et al. 1988]. During this survey, the bolter was downwind of the miner during three different cuts. For two of these cuts, the scrubber was not operating and dust concentrations measured by the intake bolter package averaged 11.23 mg/m^3. The scrubber was operating during one cut with the bolter downwind, and the intake dust concentration was 7.38 mg/m^3, resulting in a 34% reduction in dust concentration with the scrubber operating. Similarly, a 51% reduction in dust concentrations was observed for the samples located in the bolter cab. Given the small number of samples, no statistically significant difference could be calculated between the bolter-downwind dust concentrations with the scrubber off and the scrubber on.

Table C-3. Summary of roof bolter dust concentrations

a. Bolter operating location and dust levels

Place No.	Face location	Face airflow, cfm	Position of RB[*] with respect to CM[*]	Start time	End time	Scrubber status	RB intake dust, mg/m^3	RB cab dust, mg/m^3
1-1	3-heading	4,000	downwind	8:59:25	9:30:20	off	9.21	8.47
1-2	5-heading	10,140	upwind	11:02:00	11:28:10	na	0.13	0.25
2-1	1-heading	12,800	downwind	7:11:54	8:07:21	off[†]	13.25	16.04
2-2	2-heading	4,420	upwind	9:48:14	10:06:10	na	0.06	0.11
2-3	3-heading	4,360	upwind	11:11:53	11:40:24	na	0.07	0.13
3-1	4-left	10,800	downwind	8:47:30	10:00:13	on[‡]	7.38	6.06
3-2	5-heading	na	upwind	10:47:30	10:57:49	na	na	na

[*] Abbreviations: RB, roof bolter; CM, continuous miner.
[†] Miner upwind but only operating from 7:46:42 to 8:01:48.
[‡] Miner begins operating upwind at 9:17:46.

b. Impact of continuous miner on dust levels at the roof bolter

RB position and status of CM	RB intake, mg/m^3	RB cab, mg/m^3
Average dust concentration when RB upwind or with CM not operating	0.09	0.16
Average dust concentration when RB downwind of CM with scrubber off	11.23	12.26
Average dust concentration when RB downwind of CM with scrubber on	7.38	6.06
Percent dust reduction when RB downwind of CM with scrubber on	**34%**	**51%**

Table C-4 displays the mass and the percentage of quartz for miner-generated respirable dust from the immediate return for samples collected for both the scrubber-off and scrubber-on test conditions. As previously indicated, a substantial amount of rock was being cut in all of the entries. Quartz levels ranged from 12.9% to 23%, except for the last cut taken with the scrubber off. For this cut, the average quartz percentage was 7.9%. This cut was taken in the crosscut between entries 3 and 2 and was the only cut not taken up a heading. The quartz concentrations were normalized to account for differences in productivity from cut to cut. With the scrubber operating, the average quartz concentration was reduced by 14% from 1,602 μg/m^3 to 1,385 μg/m^3. There was no statistically significant difference (Wilcoxon test) between the quartz concentrations for the two test conditions.

Table C-4. Respirable quartz levels in the continuous miner return

a. Scrubber-off

Shift	Gravimetric filter number	Sampling time, min	Dust mass, mg	Avg dust conc, mg/m³	Quartz mass, µg	Quartz, %	Quartz conc, µg/m³	Shuttle cars/ min	Adjusted quartz conc*, µg/m³
1	130	130	3.887	14.95	893	23.0	3,434.62	0.24	4,007.05
	46	130	3.894	14.98	756	19.4	2,907.69	0.24	3,392.31
2	64	174	1.727	4.96	239	13.8	686.78	0.31	620.32
	1	174	0.916	2.63	124	13.5	356.32	0.31	321.84
3	61	98	1.936	9.88	160	8.3	816.33	0.35	653.06
	66	98	2.016	10.29	151	7.5	770.41	0.35	616.33

* Concentrations adjusted based upon an average of 0.28 shuttle cars per min.

b. Scrubber-on

Shift	Gravimetric filter number	Sampling time, min	Dust mass, mg	Avg dust conc, mg/m³	Quartz mass, µg	Quartz, %	Quartz conc, µg/m³	Shuttle cars/ min	Adjusted quartz conc*, µg/m³
1	560	79	1.807	11.44	324	17.9	2,050.63	0.34	1,688.76
	65	89	1.679	9.43	270	16.1	1,516.85	0.34	1,249.17
2	53	74	0.839	5.67	108	12.9	729.73	0.23	888.37
	69	74	0.736	4.97	100	13.6	675.68	0.23	822.56
3	322	63	0.654	5.19	121	18.5	960.32	0.15	1,792.59
	56	63	0.619	4.91	126	20.4	1,000.00	0.15	1,866.67

* Concentrations adjusted based upon an average of 0.28 shuttle cars per min.

c. Impact of scrubber performance

Test condition and statistics	Quartz mass, µg	Quartz conc, µg/m³	Adjusted quartz conc, µg/m³
Average quartz content with scrubber off	387	1,495.36	1,601.82
Average quartz content with scrubber on	175	1,155.54	1,384.69
Calculated p-value	-	-	*0.3939*
Percent reduction in quartz content with scrubber on	**55%**	**23%**	**14%**

Conclusions

The purpose of the study was to compare face dust levels for two cutting conditions: (1) a standard 20-ft cut with a maximum 20-ft exhaust curtain setback and no scrubber operating, and (2) a 20-ft advance with exhaust curtain setbacks of up to 40 ft while operating a flooded-bed scrubber. To assure consistent operating conditions throughout the survey, MSHA personnel assisted in the study by monitoring the mine's adherence to key dust control parameters stipulated in the ventilation plan: face airflow, scrubber airflow, water spray pressure, and number of sprays operating. Adjustments were made by mine personnel throughout the survey in an effort to maintain levels within required limits.

Difficult geologic conditions were present during the survey (steep rise in the coal seam), which complicated mining and reduced production levels. In addition, substantial cut-to-cut variability was observed in the respirable dust concentrations obtained for each test condition. Key findings and observations from the survey include:

- No statiscally significant differences (Wilcoxon test, $\alpha = 0.05$) in respirable dust levels were observed between the two operating conditions tested during this survey.

- Although not statistically significant, several sampling locations exhibited a drop in average respirable dust concentrations with the scrubber operating, as follows:

    o in the miner return, average dust concentrations decreased from 12.12 mg/m^3 to 8.75 mg/m^3

    o in the main return, average dust concentrations decreased from 6.73 mg/m^3 to 4.80 mg/m^3

    o with the bolter working downwind of the miner, dust concentrations in the intake air to the bolter decreased from 11.23 mg/m^3 to 7.38 mg/m^3

- As shown in Figure C-2, the scrubber discharge was not equipped with vanes to direct the scrubber exhaust toward the return curtain. This likely allowed a portion of the scrubber discharge air to cause turbulence and disruption in the entry airflow.

Photo by NIOSH

Figure C-2. Scrubber discharge at left rear corner of miner.

- After one cut with the scrubber operating was completed, the scrubber airflow was measured before the filter panel was cleaned. Results showed a 35% reduction in air quantity, illustrating the necessity of cleaning the filter panel after each cut.

- The differences in the types of cuts sampled for the two test conditions complicated the comparison of dust levels generated during this survey. For example, for all of the cuts with the scrubber operating, the headings were advanced between 35 and 50 ft beyond the last open crosscut. During these cuts, the miner operator was most often positioned in the heading alongside of the shuttle car being loaded. For three of the cuts with the scrubber off, the miner was taking a flush cut or the face had only been advanced 10 ft beyond the last open crosscut. In these cases, the miner operator was positioned in the last open crosscut and was not as close to the miner and shuttle car. For these cuts, the ventilating air was flowing perpendicular to the miner, rather than parallel to the miner as is the case when the heading is further developed. Also, only one crosscut was sampled during the survey and it was the initial cut for that crosscut. Consequently, only a partial line curtain was hung in the heading to provide ventilation for this initial cut.

- Substantial quantities of quartz-bearing rock were being cut in all entries. Maintenance of all dust controls is critical to minimize exposure to quartz, which was found to be over 12% for all but one cut.

The data collected for this study indicated that operation of the flooded-bed scrubber with an extended curtain setback did not result in a statistically significant difference in respirable dust concentrations when compared to standard cutting conditions. However, the available data did show that operation of the scrubber resulted in measured reductions in dust concentrations in the miner return, the main return, and in the bolter intake when located downwind of the miner. Consequently, the data suggest that continued operation of the scrubber would be beneficial.

Table C-5. Continuous-miner-generated dust concentrations (mg/m^3) for each cut

Cut No.	Cut location	Face air-flow, cfm	FBS* air-flow, cfm	CM* start time	CM stop time	Total cut time, min	No. SC*	SCs/min	CM int dust conc	CM oper dust conc	CM* RRC dust conc	CM ret dust conc	Main ret dust conc	Adj[†] CM oper dust conc	Adj[†] CM RRC dust conc	Adj[†] CM ret dust conc	Adj[†] main ret dust conc
1-1	3-heading	5,757	4,300	7:58 8:20	8:15 8:23	20	7	0.36	0.02	0.72	na	13.79	na	0.49	na	9.55	na
1-2	5-heading	5,070	0	8:54 9:00 9:15 9:39 10:17 10:39	8:57 9:11 9:24 10:06 10:36 10:41	71	17	0.24	0.06	0.31	na	23.34	7.40	0.23	na	21.16	6.67
1-3	4-heading	5,910	4,400	11:34 11:45 12:02	11:42 11:56 12:19	36	12	0.33	0.05	void[‡]	na	9.38	4.33	void[‡]	na	7.17	3.29
2-1	2-heading	6,348	0	7:47 8:20 8:53	8:02 8:50 8:55	47	15	0.32	0.02	0.04	na	6.58	5.36	0.02	na	5.63	4.58
2-2	1-heading	5,280	0	9:15 9:31 9:50 9:59 10:25	9:27 9:46 9:55 10:01 10:42	51	15	0.29	0.26	1.05	na	7.26	10.83	0.61	na	5.47	8.26
2-3	3-heading	7,560	4,350	11:52 11:58 12:08 12:23 12:42 12:46	11:53 12:03 12:17 12:34 12:43 12:58	39	9	0.23	0.10	void[‡]	na	6.43	5.84	void[‡]	na	9.04	8.19

* Abbreviations: FBS, flooded-bed scrubber; CM, continuous miner; SC, shuttle car; RRC, right rear corner.

† Adjusted dust concentrations have intake levels subtracted and are normalized for differences in face airflow and production (SCs/min).

‡ PDM sampler revealed an error code for flow rate out of range, likely indicating that the sampling hose was pinched during this cut.

Table C-5. Continuous miner-generated dust concentrations (mg/m^3) for each cut (Continued)

Cut No.	Cut location	Face air-flow, cfm	FBS* air-flow, cfm	CM* start time	CM stop time	Total cut time, min	No. SC*	SCs/ min	CM int dust conc	CM oper dust conc	CM RRC* dust conc	CM ret dust conc	Main ret dust conc	Adj† CM oper dust conc	Adj† CM RRC dust conc	Adj† CM ret dust conc	Adj† main ret dust conc
3-1	5-heading	7,398	4,100	9:18 9:27	9:23 10:16	54	8	0.15	0.03	0.09	na	4.07	1.36	0.14	na	8.81	2.9
3-2	3-left	9,200	0	11:20 11:30 11:39 11:48 11:58 12:10 12:15 12:19 12:27 12:40 12:44 12:55	11:23 11:32 11:40 11:53 12:01 12:12 12:16 12:23 12:37 12:41 12:51 12:56	40	14	0.35	0.03	0.13	na	22.27	6.56	0.12	na	25.21	7.40
Avg	-	6,565	4,288	-	-	-	-	0.28	-	-	-	-	-	-	-	-	-

* Abbreviations: FBS, flooded-bed scrubber; CM, continuous miner; SC, shuttle car; RRC, right rear corner.
† Adjusted dust concentrations have intake levels subtracted and are normalized for differences in face airflow and production (SCs/min).

Table C-6. Shuttle car loading times and dust concentrations with the scrubber off

Cut No.	SC No.	SC begin loading	SC end loading	Intake dust, mg/m^3	SC* dust, mg/m^3
1-2	2	8:54:20	8:54:42	0.04	0.51
	1	8:55:36	8:56:49	0.04	3.47
	2	9:00:00	9:03:10	0.09	3.42
	1	9:04:24	9:11:05	0.15	5.00
	2	9:15:20	9:20:03	0.08	0.32
	1	9:21:07	9:24:26	0.06	3.83
	2	9:39:03	9:44:40	0.06	0.26
	1	9:46:00	9:51:24	0.05	0.83
	2	9:53:05	9:55:56	0.04	0.11
	1	9:57:36	9:59:29	0.04	0.25
	2	10:00:02	10:05:54	0.04	0.08
	2	10:16:34	10:18:22	0.04	2.08
	1	10:19:00	10:21:45	0.04	1.37
	2	10:22:26	10:25:37	0.05	0.07
	1	10:28:36	10:30:33	0.05	0.14
	2	10:31:09	10:35:53	0.05	0.02
	1	10:38:59	10:40:40	0.05	0.21
Cut avg	-	-	-	-	**1.29**
2-1	1	7:46:42	7:47:07	0.02	0.02
	2	7:48:28	7:50:00	0.04	0.00
	1	7:51:20	7:53:13	0.05	0.06
	2	7:54:26	7:56:18	0.01	0.01
	1	7:57:28	8:00:01	0.06	0.04
	2	8:01:25	8:01:48	0.06	0.00
	1	8:20:28	8:22:18	0.02	0.08
	2	8:24:53	8:27:10	0.01	0.00
	1	8:28:53	8:32:15	0.01	0.02
	2	8:33:36	8:36:21	0.01	0.00
	1	8:38:07	8:42:07	0.01	0.01
	2	8:43:36	8:46:08	0.00	0.00
	1	8:47:26	8:50:05	0.01	0.02
	2	8:53:23	8:53:33	0.03	0.00
	2	8:55:12	8:55:34	0.03	0.00
Cut avg	-	-	-	-	**0.02**
2-2	1	9:15:18	9:16:46	0.08	0.49
	2	9:18:31	9:20:02	0.37	0.00
	1	9:21:58	9:23:20	0.05	0.01
	2	9:24:49	9:27:11	0.04	0.00
	1	9:30:40	9:32:30	0.14	0.00
	2	9:34:00	9:35:58	0.01	0.00
	1	9:37:36	9:39:59	0.09	0.00
	2	9:41:35	9:43:37	2.30	0.00

* Shuttle car (SC) dust levels have intake levels subtracted; if less than zero, zero is shown.

Table C-6. Shuttle car loading times and dust concentrations with the scrubber off (Continued)

Cut No.	SC No.	SC begin loading	SC end loading	Intake dust, mg/m^3	SC* dust, mg/m^3
2-2	1	9:45:50	9:50:16	0.29	0.00
	2	9:51:48	9:55:29	0.46	0.00
	1	9:59:12	10:01:20	0.27	0.00
	2	10:24:52	10:26:08	0.02	0.00
	1	10:27:20	10:27:35	0.01	0.00
	2	10:28:45	10:32:46	0.14	0.00
	1	10:35:30	10:41:49	0.02	0.00
Cut avg	-	-	-	-	**0.01**
3-2	2	11:20:23	11:22:34	0.01	0.01
	2	11:30:03	11:31:48	0.02	0.00
	2	11:38:38	11:40:28	0.02	0.02
	2	11:47:50	11:49:47	0.03	0.00
	1	11:52:03	11:53:26	0.03	0.25
	2	11:58:10	12:01:16	0.03	0.07
	2	12:09:44	12:11:49	0.03	0.05
	2	12:15:09	12:15:48	0.03	0.00
	2	12:19:20	12:20:43	0.03	0.00
	1	12:22:28	12:23:20	0.03	0.58
	2	12:27:13	12:29:26	0.03	0.00
	1	12:31:36	12:33:24	0.03	0.06
	2	12:35:47	12:36:31	0.03	0.11
	2	12:40:06	12:40:44	0.12	0.00
	2	12:43:56	12:48:20	0.03	0.00
	2	12:50:14	12:51:30	0.03	0.00
		12:54:40	12:55:53	0.03	0.00
Cut avg	-	-	-	-	**0.07**
Survey avg	-	-	-	-	**0.35**

* Shuttle car (SC) dust levels have intake levels subtracted; if less than zero, zero is shown.

Table C-7. Shuttle car loading times and dust concentrations with the scrubber on

Cut No.	SC No.	SC begin loading	SC end loading	Intake dust, mg/m^3	SC[*] dust, mg/m^3
1-1	1	7:58:04	7:58:35	0.04	0.23
	1	7:59:00	8:03:28	0.04	0.12
	1	8:06:24	8:07:50	0.04	0.16
	2	8:09:33	8:11:30	0.04	0.67
	1	8:12:28	8:14:54	0.04	0.12
	1	8:19:59	8:20:46	0.04	0.08
	1	8:21:57	8:22:39	0.11	0.24
Cut avg	-	-	-	-	**0.23**
1-3	1	11:33:40	11:34:30	0.04	0.20
	1	11:36:14	11:38:26	0.05	0.60
	1	11:40:00	11:41:45	0.03	2.10
	1	11:44:50	11:47:40	0.03	7.73
	1	11:50:27	11:52:27	0.05	3.64
	1	11:53:40	11:56:20	0.05	7.85
	1	12:02:36	12:03:34	0.04	1.03
	1	12:06:15	12:07:03	0.07	1.08
	1	12:08:47	12:09:51	0.05	0.31
	1	12:11:44	12:14:14	0.06	0.12
	1	12:15:29	12:16:26	0.07	0.52
	1	12:18:32	12:19:07	0.07	4.61
Cut avg	-	-	-	-	**1.33[†]**
2-3	1	11:51:46	11:52:30	0.04	0.00
	1	11:57:55	12:03:10	0.02	0.07
	2	12:08:12	12:13:04	0.06	0.00
	2	12:15:25	12:17:02	0.17	0.00
	1	12:23:05	12:23:57	0.12	0.00
	1	12:26:09	12:29:46	0.42	0.05
	2	12:31:53	12:34:06	0.09	0.00
	2	12:41:39	12:42:30	0.07	0.00
	2	12:45:34	12:48:14	0.04	0.00
	1	12:49:50	12:52:28	0.04	0.02
	1	12:54:27	12:55:49	0.03	0.00
	1	12:57:29	12:58:09	0.03	0.02
Cut avg	-	-	-	-	**0.01**
3-1	2	9:17:46	9:19:20	0.04	0.30
	1	9:20:17	9:21:42	0.05	0.25
		9:23:29	9:29:41	0.04	0.06
		9:29:58	9:32:27	0.03	0.09
	2	9:33:41	9:41:18	0.03	0.13
	1	9:42:40	9:45:20	0.03	0.02
	2	9:47:11	9:52:52	0.02	0.11

[*] Shuttle car (SC) dust levels have intake levels subtracted; if less than zero, zero is shown.
[†] Line curtain too close to face with scrubber operating until 12:05. Cut average calculated after this time.

Table C-7. Shuttle car loading times and dust concentrations with the scrubber on (Continued)

Cut No.	SC No.	SC begin loading	SC end loading	Intake dust, mg/m^3	SC* dust, mg/m^3
	2	9:54:16	10:03:47	0.03	0.22
	1	10:06:06	10:09:21	0.03	0.13
	2	10:10:59	10:16:04	0.02	0.17
Cut avg	-	-	-	-	**0.15**
Survey avg	-	-	-	-	**0.43**

* Shuttle car (SC) dust levels have intake levels subtracted; if less than zero, zero is shown.

[†] Line curtain too close to face with scrubber operating until 12:05. Cut average calculated after this time.